Dirk von Plessen

The procurement strategies for the Olympic Stadium and the Aquatic Centre for the London 2012 Olympic Games

Anchor Academic
Publishing

von Plessen, Dirk: The procurement strategies for the Olympic Stadium and the
Aquatic Centre for the London 2012 Olympic Games. Hamburg, Anchor Academic
Publishing 2015

Buch-ISBN: 978-3-95489-392-8
PDF-eBook-ISBN: 978-3-95636-351-1
Druck/Herstellung: Anchor Academic Publishing, Hamburg, 2015

Bibliografische Information der Deutschen Nationalbibliothek:
Die Deutsche Nationalbibliothek verzeichnet diese Publikation in der Deutschen
Nationalbibliografie; detaillierte bibliografische Daten sind im Internet über
http://dnb.d-nb.de abrufbar.

Bibliographical Information of the German National Library:
The German National Library lists this publication in the German National Bibliography.
Detailed bibliographic data can be found at: http://dnb.d-nb.de

All rights reserved. This publication may not be reproduced, stored in a retrieval system
or transmitted, in any form or by any means, electronic, mechanical, photocopying,
recording or otherwise, without the prior permission of the publishers.

Das Werk einschließlich aller seiner Teile ist urheberrechtlich geschützt. Jede Verwertung
außerhalb der Grenzen des Urheberrechtsgesetzes ist ohne Zustimmung des Verlages
unzulässig und strafbar. Dies gilt insbesondere für Vervielfältigungen, Übersetzungen,
Mikroverfilmungen und die Einspeicherung und Bearbeitung in elektronischen Systemen.

Die Wiedergabe von Gebrauchsnamen, Handelsnamen, Warenbezeichnungen usw. in
diesem Werk berechtigt auch ohne besondere Kennzeichnung nicht zu der Annahme,
dass solche Namen im Sinne der Warenzeichen- und Markenschutz-Gesetzgebung als frei
zu betrachten wären und daher von jedermann benutzt werden dürften.

Die Informationen in diesem Werk wurden mit Sorgfalt erarbeitet. Dennoch können
Fehler nicht vollständig ausgeschlossen werden und die Diplomica Verlag GmbH, die
Autoren oder Übersetzer übernehmen keine juristische Verantwortung oder irgendeine
Haftung für evtl. verbliebene fehlerhafte Angaben und deren Folgen.

Alle Rechte vorbehalten

© Anchor Academic Publishing, Imprint der Diplomica Verlag GmbH
Hermannstal 119k, 22119 Hamburg
http://www.diplomica-verlag.de, Hamburg 2015
Printed in Germany

Table of Contents

1 **Introduction** 1
  1.1 Background information 1
  1.2 Motivation for the study.. 3
  1.3 Scope and aim of present work 5
  1.4 Limitations of the study 6
  1.5 Significance of the study... 6
  1.6 Chapter overview 7
      1.6.1 Literature review 7
      1.6.2 Methodology 7
      1.6.3 Data & Results 7
      1.6.4 Analysis and discussion... 7
      1.6.5 Summary and Conclusions 7

2 **Literature review** 9
  2.1 Introduction... 9
  2.2 Terminology of construction procurement 10
  2.3 Client identity and characteristics.. 12
  2.4 Client needs, requirements and project objectives................................ 15
      2.4.1 Prioritisation of the client needs 16
  2.5 Project Risk... 18
  2.6 Procurement Method Selection... 20
      2.6.1 The selection of a procurement method in practice..................... 20
      2.6.2 The selection of a procurement method in theory 21
  2.7 The Selected Procurement System 25
      2.7.1 The Procurement Method.. 25
      2.7.2 The Contractor Selection... 29
      2.7.3 The Type of contract 31
  2.8 Research Framework 34

3 Methodology 35

3.1 Introduction 35

3.2 Data required 35

3.2.1 Primary data source 35
3.2.2 Secondary data sources 36

3.3 Primary Data Collection Methods 36

3.3.1 Sampling procedure 38
3.3.2 Conducting Interviews 39

3.4 Data Analysis 40

4 Data & Results 41

4.1 Introduction 41

4.2 Primary and Secondary Data Results 41

4.2.1 The Project 43
4.2.2 The Client 44
4.2.3 The Procurement Method selection 46
4.2.4 The selected Procurement System 47

5 Analysis and Discussion 51

5.1 Introduction 51

5.2 The client 51

5.3 The selection of the procurement method 52

5.4 The selected procurement system 53

5.4.1 The procurement method 53
5.4.2 The contractor selection 55
5.4.3 The chosen form of contract 56

6 Summary and Conclusion 58

7 Bibliography 61

8 Appendicies 70

Appendix 1 - Interview Schedule 70
Appendix 2 – Interview Invitation Letter 71
Appendix 3 – Quotation references - The Project 72
Appendix 4 – Quotation references - The Client 76
Appendix 5 – Quotation references - The Method Selection 86
Appendix 6 – Quotation references - The Selected Procurement System 89

List of Graphics and Tables

List of Graphics

Graphic 1-1:	The Olympic Stadium	
Graphic 1-2:	The Aquatic Centre	
Graphic 2-1:	Elements of a procurement strategy or system	
Graphic 2-2:	Tension Triangles	
Graphic 2-3:	Contractual relationship between parties in a Design & Build contract	
Graphic 2-4:	Contractual relationship between parties in a Novated Design & Build Contract	
Graphic 2-5:	Theoretical research framework	
Graphic 4-1:	Initial assessment of available procurement options for the main venues in t Olympic Park (ODA 2005)	

List of Tables

Table 2-1:	List of "key determinants" for procurement suitability and their relation to the main project criteria time, cost, and quality	
Table 2-2:	Overview of Procurement systems used for the Olympic Stadium and Aquatic Centre.	
Table 4-1:	Conceptual categories and their relevant sub-categories	

Nomenclature

ODA	–	Olympic Delivery Authority
CLM	–	CH2MHill, Laing O'Rourke and Mace
OGC	–	Office of Government Commerce
NAO	–	National Audit Office
LOCOG	–	London Organising Committee for the Olympic Games

1 Introduction

The introduction of this thesis details the background information to the subject area, my motivation for the study and the research objectives. It also defines the limitations and the significance of the study and finally, an overview of each chapter is presented.

1.1 Background information

The International Olympic Committee announced on the 6th July 2005 that the Games of the 30th Olympiad in 2012 will take place in the city of London.

Three years later, a lot of preparation work has already been done to get London ready for hosting the world's most prestigious sporting occasion. Over 192 buildings have been demolished, one million cubic metres of soil excavated, two six kilometre tunnels and 200km of cabling are completed, and most of the contractors for the new sporting facilities are appointed.

The Olympic Park will be at the centre of this large development project and spans two million square metres of the Lower Lea Valley in East London. Most of the new build venues and sporting facilities will be sited here; amongst them are the two flagship venues the Olympic Stadium and the Aquatics Centre.

At the heart of the park will be the Olympic Stadium. The brief for the stadium published by the Olympic Delivery Authority (ODA) outlines the venue as a spectacular 80,000-seat arena for the Olympic and Paralympic games, which is to be designed to host the athletics competitions and the opening and closing ceremonies. The masterplan for the stadium calls for the conversion of this structure into an athletics-led venue with capacity for 25,000 spectators after the games.

Graphic No 1-1: The Olympic Stadium for the London 2012 Olympics (Building 2008)

The Aquatic Centre, to the southeast of the park, contains two 50m pools and a 25m diving pool with seating for approximately 20,000 people. After the games, the capacity will have to be reduced to 3,500 seats and the centre's facilities made available to the local community. The building will then have to house a new health and fitness centre as well as facilities for nearby sports clubs.

Graphic No 1-2: The Aquatic Centre for the London 2012 Olympics (Building 2008)

The construction and operation of these sports facilities for the Games will be undertaken by the London Organising Committee of the Olympic Games (LOCOG). The delivery of the venues in time, within budget and to the required standard, however, is the responsibility of the Olympic Delivery Authority (ODA). The ODA is a non departmental public body and acts in essence as the delivery organisation for all the construction activity.

1.2 Motivation for the study

The rather difficult task which the ODA is facing is to deliver the above-mentioned facilities to an immovable deadline, to stay within budget, and at the same time to deliver the venues with astonishing design and build quality. These are the main criteria against which the success of this project will be measured.

Additionally, this enormous project is exposed to great political pressures and regulations. EU & National regulations for procuring the venues apply and commitments such as 'Value for Money' are to be considered by the ODA when making its procurement decisions.

With this in mind, the ODA have decided to procure the Olympic Stadium and the Aquatic Centre under the Design & Build route. For both venues the ODA has announced to use the New Engineering Contract (NEC) target cost contract.

Based on these procurement decisions and on the comments made by Tessa Jowell, the Olympics Minister, that the main schemes in the Olympic Park will not be design-led a debate has started between leading architects and the ODA. The argument is about the role of the design in the procurement of the Olympic venues and the way the ODA goes about selecting its preferred contractors.

Jack Pringle, the RIBA president, states that the use of Design & Build contracts would compromise the quality of design (Building 2006). He openly criticised the ODA strategy for the use of Design & Build contracts and said that "It is important that the process is not contractor-led, the crude old Design & Build....let's not sacrifice games excellence on the altar of the crudest form of reliable delivery" (Building 2006). Jack Pringle further argues that the ODA is acting too cautiously and by putting risk factors ahead of design at this early stage does not show a great deal of confidence (Building 2006).

In addition, Lord Rogers declares that the Design & Build contracts will lead to venues without design flair. He claims, "Every Olympic Stadium I can think of went through a design-led procurement process and I don't know why London is not doing the same. There is no proof that Design & Build contracts are cheaper in terms of value." (Sherwood 2006)

The other unpopular decision made by the ODA was to scrap the shortlist of contractors for procuring the Olympic Stadium and to go with only one bidder. The original plan for procuring the stadium was to select a preferred contractor via a short list of 3 to 6 bidders, which would help the ODA to work out the design and scope of the project.

However, the ODA decided not to go with this shortlist. Many consultants argue that this procurement decision will not only lead to a compromised design for the stadium but also to raising costs due to the absence of competition.

When looking at the procurement process for the Aquatic Centre a similar situation can be found. Despite the fact the ODA entered into a competitive dialogue with a short list of three contractors, two of them have abandoned the negotiations before any tenders could be submitted, leaving the ODA again with only one bidder for this project.

Matthew (2006) supports the above argument concerning costs by saying that it is unimaginable that London will not deliver the Olympic venues and infrastructure in time. He suggests that the real risks faced by the ODA are cost, quality and functionality and says that "Cost escalation is one of the biggest single risks. Experiences of other games and similar events indicate that as time progresses, increasing volumes of resources have been applied to overcome obstacles and costs have risen accordingly." For that reason, not having any competition in terms of price and quality seems to be a controversial decision in what is regarded one of the largest and most complex construction projects in the UK.

An auditor of the National Audit office (NAO) also shares the concerns about rising costs for the infrastructure spending in the pre-games period and says that uncertainty remains over price inflation and how much contractors will charge for the construction of the venues (NCE 2007). The Public Accounts Committee report, published in April 2008, agrees with the above and suggests that contracts should have been awarded based on effective competition between suppliers (NCE 2008).

This debate about rising costs is not unfounded under the light of the development of the total budget for the Games in the recent past. The overall budget for the Olympic Games submitted with the bid to the International Olympic Committee (IOC) was £2.4bn, back in 2004. The figure then rose to £6bn just one year after the games were awarded to London in 2005. In December 2007, Tessa Jowell has announced the final figure of £9,325bn.

These spiralling costs are also reflected in the development of the budgets for the individual projects. At the bidding stage for the Olympic Games in 2005 the Aquatic Centre was estimated at £73 million. Two years later the budget figure rose to £215 million. Balfour Beatty, as the sole bidder, then submitted costs totalling £230 million and now the cost is agreed at £242 million. The Olympic Stadium was originally priced at £280 million in London's bid document in 2005. In 2007 a final figure of £496 million was announced and only a few months later this estimate has risen to £525 million.

These debates and cost developments have paved the way for this dissertation. The scope of this study, the hypothesis and the main research objectives are outlined below.

1.3 Scope and aim of present work

It is clearly apparent from the above paragraphs that in the Pre-Olympic phase (2005-2011) the construction of the Olympic venues will be at the centre of public attention, and scrutiny. Construction industry practices will be placed under the microscope in the time leading up to the Games, especially the ODA's developed and introduced strategy for procuring the infrastructure.

The study aims to determine [hypothesis] whether the procurement strategies chosen by the ODA are the right choice for delivering the two main venues in the Olympic Park in time, on budget and to the required quality.

In order to answer this question, the approach of this study is to undertake extensive research in the subject area of construction procurement and to identify best practice in making procurement decisions for a project. In particular, the procurement strategies chosen by the ODA will be researched and their shortcomings identified.
Based on this theoretical framework, the author will be able to undertake a systematic analysis of the decisions made by the ODA to procure the two most prestigious venues in the Olympic Park. As a main part of this analysis the author will conduct semi-structured interviews with key people involved in the Olympics and with experts of

the industry. Both the literature review and the interviews will help to achieve the main research objectives of this study, which are summarised below:

- Obtain a better understanding of construction procurement and the key areas affecting the project success
- Identify best practice in selecting a procurement method and to make out the pitfalls and the shortcomings of the procurement strategies that are used for both venues
- Understand why these procurement decisions were made and identify if best practice was followed by the ODA during its procurement process and if the pitfalls of the chosen strategies were counteracted
- Determine if the chosen procurement strategies fit the client & the project

1.4 Limitations of the study

Despite the above objectives it is understood by the author that whilst the procurement strategy is an important determinant for project success, other factors, such as construction performance, client-contractor relationships, transaction cost and supply chain management will also play an important role in delivering these projects within the set parameters. Such factors could not be taken into consideration due to the university guidelines that apply to the scale of this study.

1.5 Significance of the study

The significance and importance of this study cannot be underestimated, as this study will effectively test the procurement decisions made by the ODA. The author feels that it is important to question the approach to such projects taken by the government, especially when a large amount of taxpayers' money is spent. Recent national and international examples that have experienced underperformance and as a result have wasted large amounts of public money are the Quebec Olympic Stadium, the 2004 Olympics in Athens or the Scottish Parliament building in Holyrood. It is therefore believed that this study will help to increase the understanding for the procurement decisions made by the ODA and to establish their effectiveness in helping to deliver the two main venues successfully. This has not been done before for these two projects.

1.6 Chapter overview

1.6.1 Literature review

In order to answer the research question the author will conduct a literature review of current knowledge on construction procurement in the first part of the thesis. Basic terminologies as well as current ideas in procurement are discussed and the available strategies are explained. Furthermore, the author clarifies the theory behind the procurement-strategy selection process and how the best strategy can be selected. This is then followed by a detailed review of Design & Build, competitive dialogue and target cost contracts. These are the main elements of the procurement strategies for the Olympic Stadium and the Aquatic Centre. The author will investigate their effectiveness, shortcomings and critical success factors. The findings of this review will assist to build the theoretical framework for the further research in this study.

1.6.2 Methodology

The research method used in this study is semi-structured interviews with people involved in those projects and experts of the industry. The fourth chapter gives information about why the author has decided to use this method as the instrument for gathering essential data, what the characteristics of this research method are and how the sample will help to answer the research question. The chapter also highlight why other methods have been discounted.

1.6.3 Data & Results

In the fourth chapter the author will present the results of the empirical investigation. This presentation comprises of an interview summary, which will be substantiated through selected interview participants quotes.

1.6.4 Analysis and discussion

This chapter will analyse the results from the interviews, discuss the likely consequences of procurement decisions made by the ODA. This analysis will address the main research objectives and allow for conclusion to be drawn in the final chapter.

1.6.5 Summary and Conclusions

In the last chapter a brief statement of the original problem is given. It will be concluded whether or not the procurement strategies chosen for the Olympic Stadium and the Aquatic Centre will contribute to the successful delivery of these projects.

This is followed by an answer to the essential question: "What has been achieved?" and a short discussion of the future.

2 Literature review

2.1 Introduction

Sooner or later, every client to the construction industry will be confronted with the decision of how best to procure a project in order to minimise delay in commencement and completion, minimise risk, reduce disputes and disruption, obtain value for money and deliver the project within budget constraints and to set quality standards.

Over the past years researchers have attempted to define procurement strategies depending on the type of client involved and the characteristics of the project. Masterman (2004) provides a good background on this subject and says that although the determination of an appropriate procurement strategy is not the only reason for good project performance it is a significant contributory factor to achieve a high level of project success. Bower (2003) also argues that because of the fact that risk allocation, project management requirements, design approach, and the involvement of consultants and suppliers are very much linked together with the chosen procurement strategy, the decision on which strategy to choose has a major impact on the timescale and the overall cost of a project.

Despite the fact that Walker (1995) maintains the view that it is the relationship between team members and their subsequent performance that is the most significant factor in determining project success, the majority of research suggests that the decision of what procurement strategy to use is a crucial one to make considering the effect it will have on the following stages of the project.

Based on the above, it can be said that the procurement strategy chosen by the Olympic Delivery Authority (ODA) for the two main venues in the Olympic Park will have an impact on the successful project delivery. By choosing these strategies the ODA have set the stage on which the project must play out.

Initially, the author will step back a little from the research question and establish the basic terminology of construction procurement.

2.2 Terminology of construction procurement

Procurement in construction generally embraces all those activities undertaken by a client seeking to construct or refurbish a building. The literature, however, is somewhat inconsistent in terms of the terminology it uses to describe procurement in construction. It is variously referred to as method, path, strategy or system. For that reason, it needs to be clarified what exactly is meant by these terms in the context of construction procurement and how they will be used by the author for the purpose of this work.

A typical construction project passes through a number of phases, from inception through to completion, with the inception phase commencing immediately after the client decides to construct a building. According to Masterman (2004) this inception phase is concerned with establishing a framework for successful project completion, the so called "project strategy".

The formation of a project strategy entails weighing up the benefits, risks and financial constraints that are attached to a project (JCT 2008). These will affect the choice of contractual arrangements. Masterman (2004) adds to this definition and says that developing such a project strategy is also concerned with carrying out a detailed assessment of the client's characteristics, the client's overall needs and objectives and also identifying the risks inherent in the project and the best environment to manage the design and construction.

This environment is created through the selection of an appropriate procurement strategy or procurement system, which in turn is made up of three elements. The main element is the procurement method or route, which is then complemented by the form of tender and the type and form of contract (Hanif 2007). All of these are depicted in the below Graphic No. 2-1.

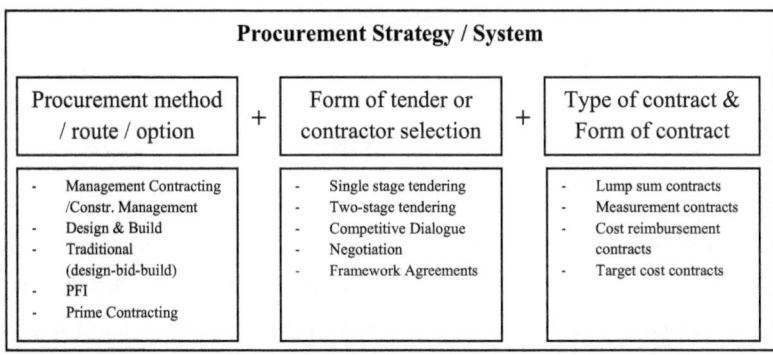

Graphic No. 2-1: Elements of a procurement strategy or system (Hanif 2007)

In order to finalise the discussion about the terminology the author would like to add that McDermott (1999) argues that the term procurement system should include not only the method used to design and construct a building but also the cultural, managerial, economic, environmental and political issues that might be generated by the decision to procure a project.

Whilst Masterman (2004) agrees that those aspects surrounding the procurement system are to be considered when making the choice of what system to use he contends that those surrounding aspects should not be part of the procurement system itself, but rather a component or sub element of the project strategy. He substantiates this argument by saying that "the sub elements do not have the ability to change the procurement systems themselves but effect the way in which the systems should be selected and used" (Masterman 2004).

The author agrees with Masterman on this point and for the purpose of this work will adopt his view and treat the procurement system itself as the method of implementing the project and the sub elements as the factors that will influence the choice of what system to use.

In the following the author will therefore examine the impact and importance of those sub elements, which are the characteristics of the client organisation, its involvement in the procurement process and the understanding of dealing with the risks inherent in a construction project. These are fundamental to the selection of the most appropriate procurement system and its elements later on in the process.

2.3 Client identity and characteristics

The client, as the sponsor of the construction process, can be defined as "the organisation, or individual, who commissions the activities necessary to implement and complete a project in order to satisfy its needs and then enters into a contract with the commissioned parties" (Masterman 2004).

Clients to the construction industry, however, exist in a very broad spectrum ranging from experienced clients who procure construction work on a frequent basis and understand the industry, to very small and inexperienced clients who are likely to only ever build once in their lifetime. Naturally, the client's level of experience and understanding will have an impact on the procurement process and different clients with different characteristics will require different approaches as to how their projects are procured.

The literature on this subject can be summarised as consentaneous and the identification of the client characteristics is widely accepted as an essential step at the beginning of the project in order to make an informed choice from all available procurement options. Already in 1984, Cherns and Bryant recognised the importance of identifying the real client and pointed out that failure to exercise caution in this respect can cause major difficulties for the project's performance.

The Scottish Parliament building in Holyrood acts a contemporary example for this and shows what can happen if a proper identification of the client is not carried out. It also acts as a reminder of the fact that clients do not always fit the chosen procurement method. The Holyrood Project was delivered 3 years later than scheduled, and with a final price tag of £414 million, ten times the originally estimated cost.

Whilst the client on this project was more than willing to become involved in the project it was characterised as a very inexperienced political client with little familiarity to either construction or the sponsorship of major construction projects (Fraser 2004). In addition, the project was not only large in size but also highly complex and of non-standard design. Despite those characteristics, the procurement method used was construction management, where most of the risk lies with the client. One of the important conclusions from the public inquiry into the failure of this project was that there was no procurement strategy document prepared and no reasoned analysis supporting the adoption of the chosen procurement method and in how far it would fit the nature of the client. As a consequence, "there was no systematic assessment of the risks implicit in the chosen procurement route and how best to manage those risks" (Auditor General of Scotland 2000, Para 3.20).

This negative example shows that the client's experience and its ability to participate in the procurement process are factors that determine the success of the chosen procurement method in terms of achieving a high quality building, keeping in control of the budget, and finishing on time.

In order to ensure that these factors can easily be established at the outset of a project researchers like (Duffy 1992), (Galbeith 1995), (Masterman 2002), or (Naphiet and Naphiet 1985) have attempted to define certain categories of clients. However, Masterman (2004) recognises that such categorisations will not be exhaustive enough to capture all various forms and subspecies of clients. He concludes that "the characteristics of clients that are most likely to be relevant to the implementation of construction projects, and more particularly to affect the choice of the most appropriate method of procurement" are the type of client and its experience.

The two main types of clients that were identified are the public and the private client. The characteristics of these two types differ mainly as a result of the source of funding and the legal framework they operate in, which in turn affects their likely approach and attitude towards project procurement.

The second important characteristic is the level of the client's experience of implementing projects in the construction industry. In support of the argument made by Masterman (2004), Moorledge (1987) also recognises this as a critical characteristic because of the effect it can have on the client's behaviour during project execution. Especially in case of megaprojects like the Olympic Games it is necessary to understand the anatomy of such projects in order to be an effective player in the project development (Flyvberg et al 2007). The main characteristics that are to be established in order to determine the level of experience of the client team are:

- Knowledge and understanding of the construction industry and its established procedures
- Regularity of involvement with the construction industry and in particular with large infrastructure projects
- The expertise of client team in the overall management of construction projects

- The ability to generate an all-inclusive brief with prioritised objectives in terms of cost, time and quality
- The ability to be constructively and consistently involved in the project from start through to completion

In addition to the above characteristics it needs to be mentioned that organisations are complex bodies, and it is therefore necessary to determine the organisational arrangements in order to refine the profile of the client. It was suggested by Handy (1985) that the organisational arrangements and culture of any client establishment is determined by its history and ownership, size, the technology it uses, its goals, objectives, environment and people.

2.4 Client needs, requirements and project objectives

Notwithstanding the importance of knowing the client's identity as a basis for choosing a suitable procurement method, it is the client himself who is able to influence the project outcome by the actions it undertakes whilst procuring a construction project.

The role of the client is an important one to be considered in the context of construction procurement because it is the client who provides the most important perspective on the project performance and whose needs must be met by the project team (Latham 1994).

Moorledge et al. (2006) found that especially for large and complex projects, like the Olympic Games, the overall outcome is often determined not only by the skill and quality of the project team, but also by the active involvement of the client body during the whole project, particularly during the initial stages of the procurement process. One of the key tasks is the client's statement of what is required or, in other words, the formulation of the project brief. An accurate and definitive statement of requirements will not only enable the choice of the most appropriate procurement system but also aid the control of all construction activities (Masterman 2004). Masterman argues that "too often, an early decision is made on building projects to appoint a team of design consultants rather than a single unbiased adviser, with the result that a procurement system is often chosen by default rather than design."

Many detailed studies and reports were produced over the last three decades, such as Bennett and Flanagan (1983), Hewitt (1985) or The University of Reading (1988) that have endeavoured to determine typical client requirements. It was found that time, cost and quality still remain the main focus of construction clients. Also Masterman in his study in 1994 attempted to prioritise the needs of clients and concluded that "there is little reason to doubt that quality, cost and time remain client's primary objectives". Importantly, he pointed out that the specific definition of these criteria will vary from client to client and project to project and it is their accurate definition that sets the stage for all subsequent activities of the project.

It is therefore considered important to briefly highlight the considerations that need to be given to these three criteria by the client and how they affect each other.

An important task the client has to fulfil in order to enable a successful project delivery is the provision of funding. Financial parameters need to be clearly set down from the start and as accurately as possible by the client. Only then can the objective, to complete the project without exceeding the available budget, be met. This involves having a clear understanding of the funds required and any budget constraints, which in turn will give guidance of how best to approach the project in terms of its procurement and what time and quality levels are realistic.

However, before the client can make its decision on what time objectives are realistic it has to carefully look at the project at hand. If the client is faced with an immovable deadline then the complexity of the design and an early contractor involvement to advise on design and construction logic must be considered. Failure to do so will evoke a direct relationship between time, cost and quality, where the cost will spiral because works have to be speeded up at a later date or the quality has to suffer because design compromises have to be made to meet the deadline.

Despite the fact that cost is often the main driver during the development of a project, quality is the main expectation (Abraham and Farrell 2003). Again, it is the client who has the greatest control over the design quality of the project, as he chooses the procurement method, appoints the consultants and sets the design standards. His leadership is required to ensure that the briefing process is carried out effectively to avoid a situation where the building produced is not fit for its purpose, despite being on time, within budget and well constructed (Wardrop 1996).

2.4.1 Prioritisation of the client needs

The aim of the procurement system is therefore to balance these primary objectives. Very often clients commence projects stating that all three variables are of the same importance. Masterman (2004) states that these three primary objectives are interrelated and conflicting in such a way that the achievement of all three at the same time, at the same level and at the same project is next to impossible. The focus of a client on only one key criterion will therefore have an effect upon the other two (Construction Excellence 2004). Such a prioritisation will normally have an impact on the choice of the most suitable procurement system. The below Graphic No. 2-2 visualises this aspect.

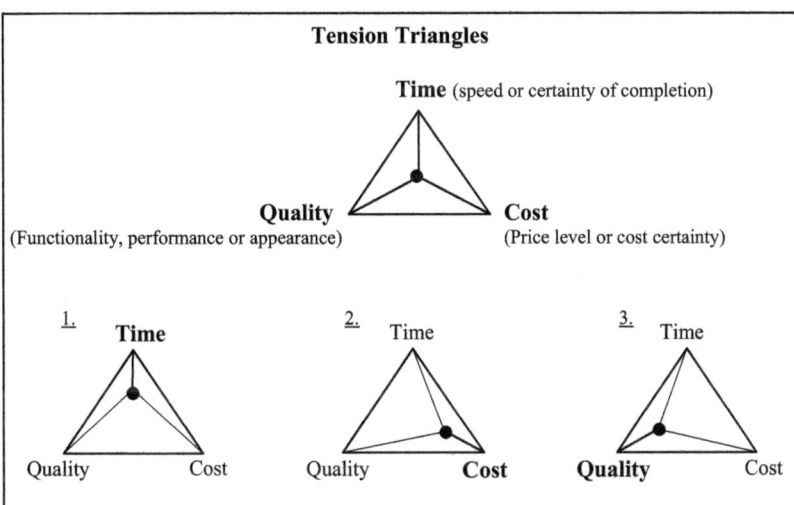

Graphic No 2-2: Tension Triangles (University of Nottingham 2008)

Walker's (1996) findings agree with the above and he adds that at least one of those main three objectives, if not two, will need to be sacrificed to some extent, and individual clients will need to weigh each of the criteria to suit their own organisation's particular circumstances and the project's technical, commercial and other characteristics.

The weighting of each of those objectives will therefore vary from project to project and from client to client, but the exact weighting needs to be accurately identified and clearly determined in order to select a suitable procurement system.

It can be summarised that researchers are unanimous on the subject of the client role and the fact that the effectiveness of a client in developing the project brief and defining and prioritising the above described main objectives will heavily influence the project outcome.

2.5 Project Risk

Projects involve commercial risk. This is one of the most significant defining characteristics of projects and project strategies. Researchers, unanimously, recognise the critical relationship between risk and cost as an important one to be understood by the client before making any procurement decisions.

Construction projects unavoidably involve certain risks and these risks can be categorised into physical works, delay and disputes, direction and supervision, damage and injury to person and property, external factors, payment, law and arbitration (Abrahamson 1984, Bruni 1985). The client will have to decide where the liability for those risks lies and to what party they are allocated. Is the risk to be with the contractor, the designer, the consultants or the client itself? It is in the nature of every contract to allocate risks and to make one party financially liable should a risk occur. This is why risk becomes a commercial concern and the chosen type of contract, within a certain procurement system, will ultimately dictate the financial implications and the degree of the commercial risk for the client.

It can therefore be said that in the same way the definition and prioritisation of the main project objectives impacts on the selection of the most appropriate procurement approach, the required allocation of risk influences the choice of what procurement method and contract structure is the best fit (Murdoch 2005).

When trying to determine the most suitable allocation of a risk it is in fact the formulation of the most suitable response to that risk that is sought. Here, the client has a range of possible options, which are the transfer, retention, avoidance or reduction of risk. The most important responses in the context of construction procurement are the transfer to the contractor or the retention of the risk by the client.

Publicly financed organisations' are generally very risk adverse and clients have been extremely shy of retaining risks. As such, risks have been transferred to the contractor by default rather than assessment.

The author feels that this is an important point to mention and a question that is to be answered when assessing the procurement decision made by the ODA for the two flagship venues. Passing every risk to the contractor is, as a general principle, not sensible, especially those risks that are difficult to assess (Murdoch 2005). This is because contractors usually include contingency amounts in their tenders as a means of pricing the uncertainty of whether or not a particular construction risk might occur and to what extent it can be controlled (Bower 2003).

For that reason, transfer of risk is not always in the best interest of the client. In particular on large infrastructure projects such as the Olympics risks should not simply be transferred by default. The construction of the venues for reasons of size and innovation alone, combined with the time constraints, attracts many risks. These can cause the costs of the whole project to overrun significantly. These risks can only be acknowledged and their impact reduced through careful identification (Flyvberg et al 2007). This will ensure that risk is allocated to the party that is best suited to manage it. Such an approach to risk will support the selection of the most suitable procurement system and will guarantee that risk is dealt with at the minimum cost to the project.

2.6 Procurement Method Selection

Now that the important role of the client during the initial stages of the project has been discussed and the basic principle of risk management put into the context of construction procurement, the author will turn to the selection of the most appropriate procurement method, which is an important element of the procurement system.

Its selection is effectively a decision-making process. This chapter is concerned with the question of how this decision should be made to suit the client's needs and objectives and the author will explore the practice and theory behind this decision-making process.

2.6.1 The selection of a procurement method in practice

It is argued by Masterman (2004) that the decision making process in practice "usually begins with very little understanding of the decision situation, a vague notion of possible solutions and very little idea of how to evaluate them and choose the most viable solution".

Moreover, he suggests that the selection in practice features a complex process that often takes considerable time to complete and is influenced by a number of parties. He adds that the way in which many clients and their advisors select their procurement method can be haphazard, ill-timed, and lacking of logic and discipline. This poses the question, how can a strategy be chosen and what are the principles of best practice in procurement method selection?

Across the literature on this subject, the necessity is stressed to carry out the selection process before a contractor is appointed. In other words, the timing is seen as important. This is to ensure that an unbiased selection of the most appropriate procurement method can be made. Whilst this seems to be a very logical approach, Masterman, in a study where he surveyed sixty-two clients, found that twenty-three percent of the clients did not follow this recommended and logical practice. Due to the fact that most of them were experienced clients he concludes that many clients do not seem to have recognised the importance of this process in the past.

The actual procurement method selection then starts with the identification and evaluation of all available options and the evaluation of their suitability for the project at hand. It is this step in the selection process that requires an independent approach to ensure that the later selection is not biased and that the project criteria are matched to the characteristics of the most suitable procurement method.

Nutt (1984), Mintzberg et al. (1976) and Hickson et al. (1986) examined more than 200 projects and clients and Masterman (2004) builds on this research by identifying five distinct categories of system identification and evaluation. These are the analytical search, consultative search, historical evaluation, intuitive evaluation and policy compliance.

Examining all the clients involved in this study he found that only one-quarter have carefully analysed and established the project criteria in order to match them with the characteristics of the most appropriate procurement method. Primarily public clients had used historical evaluation and policy compliance for this part of the selection process. Masterman (2004) suggests that this indicates an insufficient understanding of the selection principles resulting in a reliance on past experience. He follows that when such experience or policy compliance is dictating the selection process then often little consideration is given to changes in the project typology, the client's own needs or the current state of the construction market. The decision is then biased.

Considering the above findings and the importance for public clients to obtain value for money Masterman (2004) poses the question as to how this requirement can be satisfied when using a non-analytical examination or no examination at all.

The author therefore recognises this as an important point that would need to be investigated in relation to the Olympic Stadium and the Aquatic Centre. The literature on this subject confirms that, especially for temporary public clients like the ODA, the selection of a suitable procurement method should be carried out in an objective and disciplined manner.

2.6.2 The selection of a procurement method in theory

In order to make a disciplined and objective selection and to stay within the framework of the project strategy and project brief, researchers agree that several theoretical project assessment criteria should be considered by the client at the identification and evaluation stage. Some of the typical project assessment criteria that the author identified across the literature and their relation to the main 3 project objectives are listed in the below Table No 2-1.

Main project objectives expressed within the project brief	Related Assessment Criteria acting as the key determinants for the selection of the most appropriate procurement method
Cost	related Assessment Criteria: • Certainty of final cost • Value for money • Lowest possible tender
Cost & Time	related Assessment Criteria: • Ability to change design • Risk Management: Transfer, Elimination, Avoidance, Minimisation of risk • Complexity of project
Time	related Assessment Criteria: • Minimum design and construction period • Certainty of completion date • Early Start on site
Quality	related Assessment Criteria: • Innovation • Appearance • Functionality
General	Related Assessment Criteria: • Responsibility and involvement • Accountability • Management (i.e. single point of contact)

Table No 2-1: List of "key determinants" for procurement suitability and their relation to the main project criteria time, cost, and quality

Many theoretical models have been established based on the above criteria to aid the selection process. Masterman (2004), however, suggests that the early models from HM Treasury's Central Unit on Purchasing (1992), Mohsini and Davidson (1989) or Birrell (1992) are only very basic means and can only act as a 'primer for discussion'.

Nowadays, the selection process has become more and more complex as a result of the increasing technical complexity of projects, the need for speedy start and completion, and the continuing proliferation of different methods. This has led to the development of more sophisticated and systematic approaches of system selection and researchers over the years have attempted to develop more systematic approaches for the procurement method selection (Masterman 2004).

Examples are Skitmore and Marsden (1988), Franks (1990), Bennett and Grice (1990), the Construction Round Table (1995), (Chan et al. 2001), (Cheung et al. 2001), (Luu et al. 2003) or the model included in the OGC Procurement guide (2007).

Masterman (2004) carried out a trial based on the model of Skitmore & Marsden to establish which procurement system should have theoretically been used on the projects that were part of his study. He found that two-thirds of the choices he examined were to some degree inappropriate and did not match with the outcome of the theoretically assessment.

Although this finding was heavily conditioned by Masterman he concluded that a large amount of clients adopted an incorrect approach. The method they chose was not the best fit to their main project objectives.

Despite the objections raised by Moorledge et al (2006), who argue that there is no such thing as universal "best practice" in construction procurement it is suggested by Masterman (2004) that the above models are definitely helpful for identifying the procurement systems that would not be suitable for a particular project or client. Their application can assist in narrowing down the choice of systems that clients should consider. This view is shared with other researchers who confirm that despite the fact that these theoretical models cannot make the decision for the client they can act as a very good tool of assistance when evaluating the available procurement methods. This is because they give a good indication of the weaknesses and strengths of all available methods.

Avoidance of using such tools means that clients have to rely on their own expertise or the knowledge of external consultants and have to trust that such sources have a full understanding of the characteristics of all available options. Masterman (2004) and Skitmore and Marsden (1988) highlighted that this is not the case and found that the knowledge of many consultants on the available procurement methods is incomplete. In addition, Galbraith (1995) suggests that especially experienced clients tend to drive the process with their knowledge about what has worked in the past and are therefore influenced more by experience rather than by project specific factors.

Developing an informed strategy, especially for large projects like the Olympic venues, is therefore dependent on knowledgeable and impartial advice before a particular procurement method is chosen. This will ensure that an objective selection is made. Hereby, theoretical models are proven to be of great assistance.

The procurement system selection, however, does not only include the selection of the most suitable procurement method but also the selection of the contractor to carry out the work and the subsequent formulation of the contractual relationship with this contractor. The elements of the procurement system that were used by the ODA for the two venues will therefore be discussed in the following chapter.

2.7 The Selected Procurement System

The author mentioned at the beginning of this literature review that any procurement system for a project comprises of three elements, which are the method of procurement, the form of tender and the type of contract. The following chapter will therefore explore and discuss the elements that were chosen for the two flagship venues. In the below Table No 2-2 an overview of these elements is given.

Venue	Procurement Method	Form of tender	Form & Type of contract
Olympic Stadium	Design & Build	Negotiated	NEC 3 Target cost
Aquatic Centre	Novated Design & Build	Competitive Dialogue	NEC 3 Target cost

Table No 2-2: Overview of Procurement systems used for the Olympic Stadium and Aquatic Centre

It should be noted that the aim of this chapter is not to discuss the advantages of the above elements but to highlight the points that could jeopardise the successful completion of a project when using these particular elements in a procurement system. This discussion will assist to identify the associated risks and how they can be counteracted.

2.7.1 The procurement method

Adam (1999) has shown that the majority of clients regard Design & Build as the optimum route to obtain value for money. This popularity, according to Smit (1995), arises from the perceived ability to integrate the design and construction process. Saxon (2000) argues that such integration offers better performance in terms of time and cost. Masterman (2004) also allocates Design & Build into the category of integrated procurement methods because the two basic elements of design and construction of a building become the responsibility of one organisation. He claims that Design & Build procurement method is the most famous method of this category, compared to its variants Novated Design & Build, Develop & Construct, Package Deal and Turnkey.

The philosophy of Design & Build can be defined as "a construction procurement method where the contractor offers to undertake the entire design and construction of a project" (Cox and Townsend 1998).

The relationships between the parties involved in a Design & Build scenario are illustrated below.

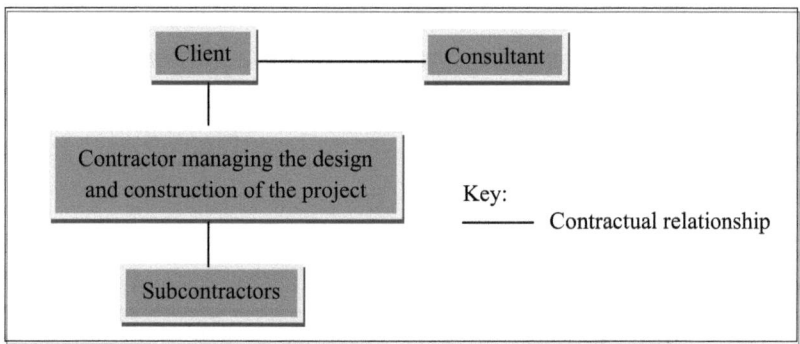

Graphic No 2-3: Contractual relationship between parties in a Design & Build contract

Masterman (2004) notes that there is a general agreement between researchers that the use of the Design & Build procurement route can be associated with a reduction of the overall project time, single point responsibility and cost savings compared to other procurement routes. This is due to the method's ability to overlap and integrate the design and construction phases, which in turn results in improved communications amongst the project team and improves the buildability of the project. However, for this to be achieved certain pitfalls need to be acknowledged, otherwise the benefits of this method will not be seen.

The first potential pitfall that can be identified is the briefing stage. As discussed earlier in this review a clearly presented and sufficiently comprehensive brief will allow tenderers to develop an accurate bid, but at the same time leaving enough freedom for innovation and to apply their design, technical and managerial expertise.

Masterman (2004) and Moorledge et al. (2006) suggest that research has shown that clients are often not able to provide such a clear and comprehensive brief and often lack sufficient information as to what their requirements are. Where the specification is not well developed, there is the risk that the quality, design and performance of the completed facility may be compromised. OGC (2007) therefore state that such specification needs to be given a lot of attention to ensure that the outcome of the project can meet the requirements.

The briefing stage itself is directly linked to the certainty of the final project cost and the likelihood of variations or compensation events. Cost certainty can only be achieved if the client's requirements are clear and properly communicated at the briefing stage and not subject to changes during the construction period (Mosey 1998). Wardrop (1996) is in support of this view and raises the concern that "All projects change and develop, for any number of valid reasons. Even the simplest of design tasks, like fitting a new kitchen or bathroom, will result in many changes as the project develops, no matter how carefully it has been planned". Murdoch et al (2005) even suggest that a client who wishes to maintain the right to change the requirements during the design and construction process should not use Design & Build at all.

On the subject of making changes, Masterman (2004) is in agreement with the above researchers but points out that, these days, most organisations are aware of the need for flexibility in terms of dealing with variations. This is achieved by the inclusion of an accurate breakdown of the contract sum in the contract documentation in order to easily accommodate the evaluation of additions or omissions.

Murdoch et al. (2005) also argue that the single point responsibility and a fixed price do not come cheap due to the fact the contractor carries all or at least most of the risk under this procurement method. The principle of risk allocation was described earlier in this review and Design & Build is often accompanied by contractual arrangements where all or most of the risk is allocated to the contractor, who is not necessarily the best party to manage it (Abrahams and Farrell 2003). The level of risk that is allocated to the contractor and the complexity of the project are therefore important factors that will have an impact on the cost of the project. Consequently, on high risk projects risk allocation is to be considered strongly in terms of cost viability.

Another reported weakness of Design & Build is the conflict between aesthetic quality and ease of fabrication. Most literature does not identify aesthetics of the finished building as a benefit of this procurement method, but argues that it is most suitable for simple and uncomplicated projects. This is owing to the fact that it is assumed that the contractor will do what is necessary to increase the buildability of the project rather than aesthetics under the lump sum contract for reasons of cost. In addition, the general view is that there is little control over the design and the aesthetics of the building by the client during the design and construction phase.

Masterman (2004) however does not support this view and claims that many complicated and prestigious buildings were successfully completed using this method. He concludes by saying that the risk of obtaining a crude design solution is only given if the client is unclear about its requirements and the contractor selection is not carried out correctly. Also Gidado and Shamsaddin (2004) conclude that the perception of poor quality in the Design & Build procurement method is outdated and baseless. Masterman (2004) concludes that aesthetically challenging buildings are achievable but require a highly disciplined client in terms of managing the project. Moorledge et al (2002) are in agreement with this and add that the approach to construction procurement must be dynamic rather than static. The "one size fits all" approach will be unsatisfactory in most of the cases, as projects become more complex. They suggest that generic procurement processes must evolve and develop in order to fit the very specific requirements of many clients.

What can be gained from this discussion is that Design & Build can be used for any project where it makes sense to combine the responsibility for both aspects of the building process. An "off the shelf" approach however is unlikely to be successful, especially on large, high-risk and complex projects.

The procurement method used for the Aquatic Centre is Novated Design & Build, which is a variant of the ordinary Design & Build method. The relationships between the parties under Novated Design & Build are illustrated below.

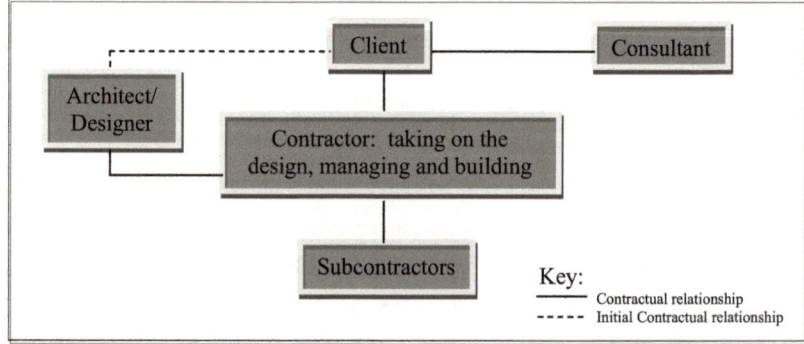

Graphic No 2-4: Contractual relationship between parties in a Novated Design & Build Contract

Murdoch et al (2005) says that although it appears to be a form of Design & Build, it is in fact leaning closer towards the traditional method of procurement and comes with a significant shift of risk apportionment.

This method is also seen to be more quality orientated, as clients can control the initial stages of the design process. The philosophy of novated Design & Build can be defined as "a construction procurement method where the client initially employs the consultant team, to carry out design and documentation, to the extent that the client's needs and intent are clearly identified and documented. The agreement with the design team is then novated to the contractor who takes responsibility for the project including the design" (Siddiqui 1996).

The two biggest pitfalls of this variant that have been identified in the literature are the "forced" working relationship between the contractor and the designer and that contractors will assess the novated design and price it as a risk. Hackett (1998) argues that the novated system "flies in the face of the philosophy of Design & Build" because contractors are required to inherit an almost complete design and are asked to take on all the risk for it. He says this is like "risk dumping" and avoids that risks being placed with the party best able to manage it. He concludes that this system does seem to mirror traditional methods yet passing the risk completely onto the contractor. Such risk allocation will show in the price for the project.

2.7.2 The contractor selection

The contractor selection, i.e. finding the right team for the successful completion of the project, is the second element in the formation of the procurement system.

The client's main objectives in this process are to obtain a fair price for the work and to enter into an agreement with a contractor who has the technical skill, resources and financial backing to complete the project within the required time, cost and quality standards (Bower 2003). The contractor selection precedes the formation of a contract, which forms the legal basis to the rights, liabilities and duties of the parties involved in the project.

The contractor selection methods available to the public client are "Open", "Restricted", "Negotiated" and "Competitive Dialogue". There are clear rules and guidelines for when each of these can be used.

For the Olympic Stadium the "Negotiated" procedure was chosen. When using this method a report must be submitted to the EU commission justifying its use. Contracting authorities are permitted to use this procedure only where the nature of the works that is procured or the risks related to them are such that prior overall pricing is not possible (Freshfields 2005).

The process consists of two rounds. Firstly, all interested contractors can submit an expression of interest. In the second round the contracting authority chooses three or more companies with whom to negotiate, provided that there are sufficient acceptable firms arising from the first round. A contracting authority is able to negotiate directly with one single contractor, but regulations are very restrictive to justify this (Moorledge et al. 2006).

The disadvantage of the procedure is obvious and researchers are unanimous on this. Due to the lack of competition the question of value for money arises.

For the Aquatic Centre the "Competitive Dialogue" procedure was used.
Before the public contract regulations introduced the procedure of "Competitive Dialogue" for the award of "particularly complex contract" contracts were mainly awarded under the "Negotiated" procedure (Black 2006). Black (2006) sees the Competitive Dialogue as being available in circumstances that prevent the usage of the "Negotiated" procedure. Freshfields (2005) even have the understanding that the use of the dialogue procedure is encouraged by the EU commission and is preferred over the use of the "Negotiated" procedure, which is, according to the commission only applicable to a very narrow range of contracts.

The intended use of "Competitive Dialogue" is for particularly complex contracts, where authorities may know what their needs are, but may be unable to decide the most appropriate technical, legal or financial solution. Contracting authorities may also wish to encourage innovative solutions, or may be unable to objectively assess what the market has on offer (Moorledge et al. 2006).

The process is split into two principal phases. The first phase, following a pre-qualification procedure, allows the authority to discuss the requirements and the solutions in greater detail, from this two to three bidders are considered for the second phase. In the second phase commercial matters are considered and the surviving bidders submit their final tenders on the basis of the solutions they have presented. At the end of the dialogue, in phase two, a contract is awarded to the economically most advantageous tender.

It is stressed by Black (2006) that this procedure must not distort competition. He concludes that the biggest risk of this process is that not enough bidders are left for there to be a genuine competition.

2.7.3 The type of contract

The type of contract is the last step in the formation of the procurement strategy and will be looked at now.

Contracts for construction services generally have the purpose to allocate the responsibility for coordinating the design and construction process, how reward is to be assessed and how risk is to be apportioned. This is true for whatever procurement method is used on a project.

On this project, for both venues the NEC 3 target cost contract option has been selected. Under a target cost contract the contractor is reimbursed for the cost of the works plus a fixed percentage fee, which includes management cost, overheads and profit. The competitive element on this type of contract should be derived from competition between contractors on the basis of the percentage addition for profit and overheads.

In general, cost and time targets are fixed unless a compensation event occurs that has an effect on cost or programme, in which case the targets will be adjusted.

Financial effects of cost overruns can be shared between the client and the contractor, and even the supply chain. This is referred to as the gain/pain share mechanism. Once the project is complete, such mechanisms come into force and all payments made to the contractor are compared to the latest agreed target cost. The gain/pain share ratio is an important contractual tool in this form of contract as it defines the risk for the client and limits its exposure to cost risk.

The client's relatively great exposure to cost risk is known to be the main downsides of this form of contract. To control this risk it is essential to set a realistic and correct target cost and to establish a clear link between the target and the actual cost. These measures, together with the gain/pain share mechanism, will be most important to control the client's exposure to cost risk.

Setting a realistic and correct target cost is important and difficult at the same time. If it is set too low it means the contractor will try to recover costs by compensation events and by adjusting the target cost. If the target cost is too high than the contractor might not be motivated enough to work efficiently. Moorledge et al (2006) and Eggleston (2006) agree with this and found that it is in the contractor's interest to win the contract with the highest possible target price. Also Turner (1996) says that "the fee is based on an agreed target estimated for the prime cost of the work and the relationship of the actual with the estimated prime cost affects the fee. Setting the target correctly is the difficult part. If the target is set too high the incentive for efficiency is reduced, if it is set too low the fee is adversely affected. In theory the system is attractive but in practice it can be difficult to achieve the objective of an incentive but with a large prime cost element."

Rawlinson (2007) therefore argues that target cost contracts are best used on well defined projects. Eggleston (2006) agrees with this and comments that for the application of the NEC target cost contract there must be a reasonable definition of the client's requirements at tender stage to enable an accurate setting of the target price. This is because the NEC3 target cost contract entitles the contractor to compensation events. Projects where the design is not sufficiently developed before the target cost is agreed run the risk that changes to the design are extremely likely. This will result in compensation events and adjustments to the target price.

Another risk pointed out by Rawlinson (2007) and Eggleston (2006) is that the target price in itself can have a complex structure, which creates the possibility that the gain/pain share mechanisms might not be understood by all parties and that control over the separation of target and actual cost is lost before completion of the project. This risk can be eliminated through appropriate contract administration and the establishment of a clear link between target and actual cost.

It can therefore be concluded that on projects where the client has a well-defined scope a target cost contract can be viewed as a cooperative approach to procurement based on risk sharing as an incentive. However, for this type of contract to unfold its benefits a realistic target cost is to be set and the relationship between target and actual cost is to be well-understood and administered. If it is used correctly, Rawlinson (2007) confirms that target cost contracts can deliver projects on time and to budget.

2.8 Research Framework

This literature review has identified a number of key areas in construction procurement that can have a significant influence on the project success and it has highlighted the areas that are to be investigated in relation to the procurement of the Olympic Stadium and the Aquatic Centre. Consequently, these key areas deliver the theoretical research framework for the further research in this study and they have been summarised by the author in the below graphic.

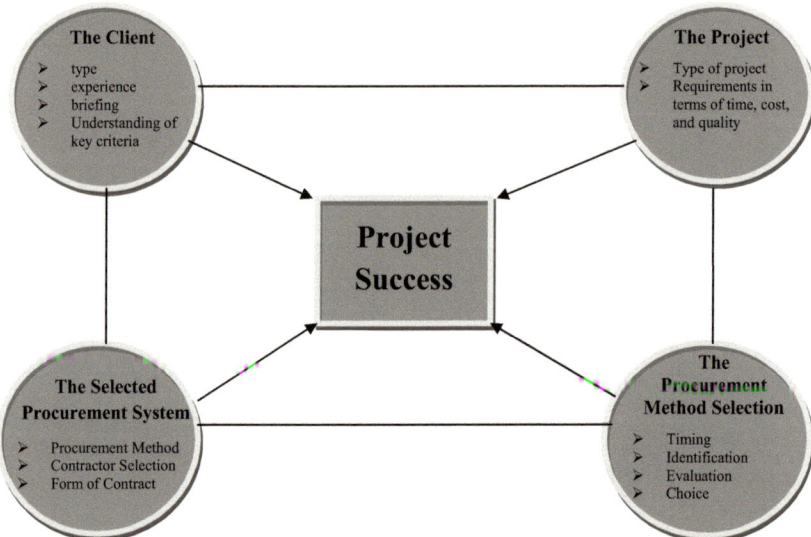

Graphic No 2-5: Theoretical research framework

This framework was instrumental in informing the scope and detail of the author's empirical investigation. The interview questions will concentrate on these areas.
The research methodology will be dealt with in more detail in the following chapter.

3 Methodology

3.1 Introduction

The purpose of this chapter on methodology is to provide the essential linkage between the research objectives, the theoretical framework gained from the literature review and the empirical investigation.

More specifically, it includes the rationale for the appropriate choice of a particular tool for the data collection. This will help to ensure that the findings of the empirical investigation provide results exhibiting a high degree of validity and reliability to "paint" a credible picture of the procurement activities undertaken by the ODA.

This rationale also considers what alternative methodological tools might have been employed, together with their advantages and limitations for this study. The focus of this chapter therefore looks at the nature of the data required, data collection methods, sampling procedures and data analysis methods.

3.2 Data required

Both primary and secondary data will be required for analysis to ensure that justifiable conclusions can be drawn and also to address the research question.

Primary data is personally collected data, either through interviews, questionnaires, focus groups, or observations. Secondary data, on the contrary, is data collected or analysed by someone other than the researcher and is presented in written form for use by others.

3.2.1 Primary data source

Primary data sources that were considered as part of the empirical investigation for this study have included the following:

- Data collected from semi-structured interviews conducted with people involved in the procurement activities for the Olympic Games in London in 2012
- Data collected from semi-structured interviews conducted with experts in the industry and specialist in construction procurement

3.2.2 Secondary data sources

Secondary data sources used in the empirical investigation have included the following:

- Procurement strategy for the Olympic venues, officially released by the ODA
- Articles, reports, and commentaries from reputable journals

3.3 Primary Data Collection Methods

A profound impact on the choice of the data collection method used in the empirical investigation is the understanding of the difference between quantitative and qualitative research methods.

De Vaus (2002) states that quantitative methodologies have their strengths in standardisation, reliability, and measurement, usually through surveys with larger sample sizes.

Qualitative methodologies on the other hand are best suited where the richness of data and the ability to understand and explore perceptions, behaviours and motivations of individuals are of importance to the researcher.

Patton (1990) describes qualitative research as a naturalistic approach that seeks to understand phenomena in context-specific settings. It involves documenting real events, recording what people say, observing specific behaviours, or studying written documents (Neuman 2000). Qualitative methods can be used to gain new perspectives on things about which much is already known, or to gain more in-depth information that may be difficult to be obtained quantitatively (Strauss and Corbin 1990).

As a result of the above findings, qualitative, primary data collection methods by means of questionnaires or interviews were initially included as possible options for the author's own empirical investigation. Quantitative methods were deemed inappropriate.

The decision was then made to select face-to-face interviews as the most suitable data collection method that best addresses the research objectives of this study. The deliberate choice not to use telephone, mail or internet surveys, using a predefined questionnaire, was made for the following reasons:

- Due to the complexity of the information required it was unlikely that respondents would have returned the questionnaire with good quality information. It was deemed unacceptable to expect respondents to write pages of text.
- Responses in questionnaires cannot be further explored if ambiguous or if an interesting fact is brought up.
- Due to the sensitive nature of this subject it was felt by the author that data collection via a questionnaire is not appropriate.
- Ambiguity in wording of questions might cause the need for further clarifications to enable the respondent to fully understand the questions and provide accurate and concise responses.
- To improve the credibility and quality of the data it was important to be very selective in choosing the individuals taking part in this study. A large sample would not have helped to gain a higher degree of data credibility.

The author's advantage of having access to people involved in the procurement activities of the ODA creates an ideal opportunity to conduct face-to-face interviews enabling the researcher to explore the interviewees responses in a personal dialogue and to obtain in-depth and credible information.

When conducting interviews researchers have the choice of selecting structured or semi-structured interviews. Esterberg (2002) suggests that most qualitative researchers should choose semi-structured interviews over structured interviews because structured interviews can give too much control to the interviewer who controls what questions are asked and how they are worded. He argues that this level of control and structure in the questions and answers can also overlook issues that are more important to the interviewee.

Esterberg (2002) further states that semi-structured interviews are far less rigid than structured interviews and ultimately address the goal of exploring a topic more openly, thus allowing interviewees to express their opinions and ideas in their own words.

For these reasons the author has decided to use semi-structured interviews. Despite the fact that the questions were designed to gain specific information they also encourage the respondents to express their views and perceptions freely and add points outside the scope of the questions when it is felt necessary by the respondent.

3.3.1 Sampling procedure

Flick (1998) further states that qualitative researchers determine the way in which people to be studied are selected on the basis of their relevance to the research topic rather than their representativeness of a population. Neuman (2000) supports this view and confirms that qualitative research should focus on how the sample can clarify and deepen understanding in a specific context. According to Patton (1990) purposive sampling is the dominant strategy in qualitative research. Here, sampling is conducted with a purpose in mind and a target sample can quickly be selected. In addition, the respondents are chosen because they have particular features or characteristics which will enable detailed exploration of the research objectives (Trochim 2000). From several purposive sampling methods[1] "expert sampling" was chosen to be most suitable for collecting the primary data for this study. The reasons are as follows:

- The specific information required for this study needed to be provided by a sample of experts with sufficient knowledge of the research topic
- The number of people interviewed was less important than the criteria used to select them
- The size of the sample was not an important factor as long as the information gathered was sufficiently rich in detail to be able to clarify and deepen the understanding of the issues and thus facilitate accurate and credible data for further analysis

The author feels that it is important to note that whilst this sampling procedure has the risk that particular views and opinions are overweighted and that sampling bias is introduced into the research being undertaken, it still provides the required richness and desired quality of data needed to conduct a successful qualitative research project (Patton 1990, Trochim 2000, De Vaus 2002).

---

[1] such as "modal instance sampling", "quota sampling", "heterogenic sampling", "snowball sampling" or "expert sampling"

3.3.2 Conducting interviews

The ultimate objective of the interviews was to obtain primary, qualitative data in the form of in-depth and detailed verbal responses from the respondents. It was hoped that their responses would be able to address the key areas identified in the literature review and the ultimate research objectives.

The experts were primarily chosen based on their knowledge of and experience in construction procurement, their involvement in the procurement of the Olympic venues and of large construction projects in general at a very senior level.

The importance of selecting highly knowledgeable senior managers stemmed from the fact that their answers would be well informed with a high degree of accuracy and credibility.

It was possible to conduct six interviews in total. Out of these, three people are directly involved in the procurement of Olympic venues, one respondent is part of the ODA procurement team and two respondents can be classed as experts of the construction industry with regard to procurement of large scale projects.

It was a natural requirement to formulate an interview schedule. For this, the literature review was instrumental in informing the scope and detail of the interview questions. A sample of the interview schedule is attached in **Appendix 1**. Together with these questions a letter, as shown in **Appendix 2**, was issued to all respondents explaining the purpose of this study. It was issued a few days before the interview took place. This letter also confirmed that the research is conforming to Kingston University's Guidance and Procedures for Undertaking Research Involving Human Subjects.

3.4 Data Analysis

The six semi-structured interviews provided many pages of typed interview transcripts. These needed to be studied in depth, organised and interpreted.

Despite the fact that Estberg (2002) and Neuman (2000) recognise that there is no single or "right" way for organising and analysing qualitative data, be it primary or secondary, it is suggested that conceptualisation is an effective way to examine qualitative data. The author has therefore decided to use conceptualisation for presenting and making sense of the data. Through this process data can be analysed by organising it into categories on the basis of themes or concepts (Neuman 2000).

For that reason, it was the author's goal to form clear and descriptive categories under which the results of the interviews could be outlined and analysed thereafter. Hereby, the research framework from the literature review was beneficial. The key areas within this framework were used to provide these conceptual categories.

The following chapter will outline the results of the empirical investigation.

4 Data & Results

4.1 Introduction

In the following chapter the author will present the primary data that has been collected during the interviews and also the secondary data referred to under section 3.2.2.

4.2 Primary and Secondary Data Results

The collected primary data was organised in two stages. In the first stage, almost all answers from the respondents were grouped into one of the four conceptual categories, irrespective of their initial perceived relevance to the main research objectives.

In a second stage, these answers were then re-examined in depth to determine their relevance to this study. As a result of this re-examination, key phrases and answers from respondents were identified and grouped into relevant sub-categories. The list of conceptual categories and relevant sub-categories can be found in **Table 4-1**. All relevant quotes from the participants for each of these categories can be found in **Appendix 3 to 6**.

In the following the author will present the results of the empirical investigation based on these categories. This presentation comprises of an interview summary, which will be substantiated through selected interview participants quotes. In addition, secondary data will be referred to in order to provide complementary support and to improve the credibility of the findings of the primary data collection.

		Conceptual Categories and their relevant sub-categories	
		Conceptual Categories	Relevant sub-categories
Related Interview Questions	All	The Project	<ul><li>Type of project</li><li>Requirements in terms of time, cost and quality</li></ul>
	1 to 6	The client	<ul><li>Type of client & experience</li><li>Project Briefing, understanding and the prioritisation of needs</li><li>Understanding of the risks involved and the attitude towards risk allocation</li></ul>
	7	The Procurement Method Selection	<ul><li>The procurement process & the evaluation of the procurement options</li></ul>
	8 to 12	The selected Procurement System	<ul><li>Procurement methods<ul><li>Design & Build and Novated Design & Build</li></ul></li><li>Contractor Selection<ul><li>Negotiation and Competitive Dialogue</li></ul></li><li>Form of contract<ul><li>Target cost contract</li></ul></li></ul>

Table No 4-1: Conceptual categories and their relevant sub-categories

4.2.1 The Project

There is a clear understanding amongst the respondents that the Olympic Stadium and the Aquatic Centre are exposed to great political & industry pressures. The ODA has an appreciation of the fact that delivering a multi venue park on this scale within a fixed timetable is a task unprecedented in the history of UK construction. It was pointed out by one of the respondents that this task consists of procuring two and a half thousand contracts, of which a hundred and fifty are major construction contracts. Out of these, the Olympic Stadium and the Aquatic Centre are the two signature projects. Both are very complex buildings. In addition, there is a huge amount of public interest and many stakeholders are involved who have vested interests in these projects. Considering and managing all of these expectations will be one of the major tasks during the procurement stages, as one of the experts' highlights:

> *"Every time we kick off a procurement ...we have actually got to consider the impact on the local community. It includes the diversity of what we are doing, the sustainability of what we are doing in terms of environment, economic particularly in social impacts, so the issues we have to consider are wide and varied."*

To deliver the venues successfully they have to be built on time, within the agreed budget and to the right quality. The deadline 2012 is fixed and it is a target that cannot move. Time, therefore, is the ultimate requirement. The clients funding package is made up of grants from the National Lottery, the London Development Agency, the Department for Culture, Media and Sport and also council tax precepts from the Greater London Authority (ODA 2007). Putting it briefly, public money is spent on a large scale and therefore transparency is a key element, which is to be ensured across all stages of the project.

The visual characteristics of both venues also have a major part to play. The Aquatic Centre rests on its aesthetic appeal, so does the Olympic Stadium, which is the centrepiece of the Games. According to most of the respondents the ODA realises the fact that people's perception of the Games all around the world will be heavily influenced by the design quality of these two projects.

> *"I mean at the end of the day the public purse is going to have to pay for it so we can't be wanton with our expenditure." "But we also have to build something that makes a statement about London and about the UK."*

4.2.2 The Client

The ODA, as the client of these projects, is a public organisation that was formed very quickly. At the beginning, it was suggested that the ODA did not appear to be a particularly educated client. However, experienced advisors were brought in to assist the ODA in making strategic decisions and very senior and experienced professionals from within the construction industry were hired to transform the ODA into an informed client with a clear understanding of leadership. In addition, a lot of the responsibility was divested to their delivery partner CLM[2]. This partner was appointed to provide the management and technical capability and systems to manage the planning, design, procurement and the delivery of the venues on behalf of the ODA. This leaves the ODA to be a thin construction client who concentrates mainly on strategic decision making and cost management on behalf of the government.

> "...I think genuinely, Her Majesty's Government has gone out there and bought the best people they could find and are available within construction."

Despite the above, it is suggested by the respondents that bureaucracy and long-winded decision making processes are evident as in so many government organisations due to auditability guidelines and the amount of stakeholders involved.

It also apparent that owing to the fact that initially the ODA did not appear to be an educated client the above mentioned advisors have very much influenced the clients thinking in terms of what is required to build these two venues. The fact that time is the most important criteria did not require any advice, but to get the balance right in terms of the three key elements of time, cost and quality the client was very much supported by the consultancy firms that were involved at the early stages of the project and later by their delivery partner CLM. It was confirmed that a huge amount of consideration went into getting this balance right and that clear a business case was in place for both venues prior to making any procurement decisions. A respondent who is involved in the procurement of these venues said that:

> "The teams here are very big and for good reason because prior to kicking off a procurement undertaking you need to understand the issues of time, cost and quality at the simplest level."

---

[2] **CH2MHill, Laing O'Rourke, Mace**

In addition, both the respondents and the ODA's procurement policy confirm that the OGC gateway process is followed to ensure that there is an understanding of these key criteria from the very start and to achieve the necessary approvals from government.

Public sector clients are generally risk averse and the ODA does not appear to be an exception to this rule. The initial plan on the Olympic Stadium was to transfer all the risk to the contractor by choosing the Design & Build procurement route together with a fixed price lump sum. This shows that initially the client did not fully appreciate the risks involved in this project. One respondent points out that:

> *"...with very large infrastructure projects, whereby, essentially it is never been done before, so to try to get a lump sum price from a contractor when the design does not exist or it exists to such an extent that it is meaningless, either means, that the contractor is going to charge an enormous amount of premium for taking the risk or he is going to leave you with the risk and is not going to take it at all..."*

However, the client's attitude towards risk allocation changed over time. The respondents suggested that this was a development supported by the construction market, concerns over cost and a lack of interest of the supply chain to submit a bid for the stadium. This change in direction is supported by the participants quote below:

> *"...you got to think about it in that it's in the shadow of Wembley, there were so many issues with Wembley with managing the project, a lot of contractors said why on earth would we divert time and resources and risk into a project like that when there is much other valuable work going on out there"*

> *"I think, because the design and build cost were coming in so high because McAlpine put so much contingency in there... and CLM have said, as a professional team, lets us take on some of that responsibility so you actually changed the type of contract into a target cost, and we are now trying to reduce that contingency because we are going to manage it."*

The Aquatic Centre, on the other hand, was commenced much earlier in terms of procurement. It was suggested that there has been an extremely thorough briefing process prior to initiating the procurement of the works and that the risks and requirements of this project were very well understood by the ODA.

4.2.3 The procurement method selection

The procurement process itself was commenced very early by the ODA and available procurement options had been evaluated together with an experienced consultancy in a series of workshops before the ODA's delivery partner CLM was appointed. A draft procurement strategy was prepared as an initial assessment of the available procurement methods for the main venues in the Olympic Park. This initial assessment was issued for discussion and review with the main stakeholders. The below Graphics 4-1 shows this assessment.

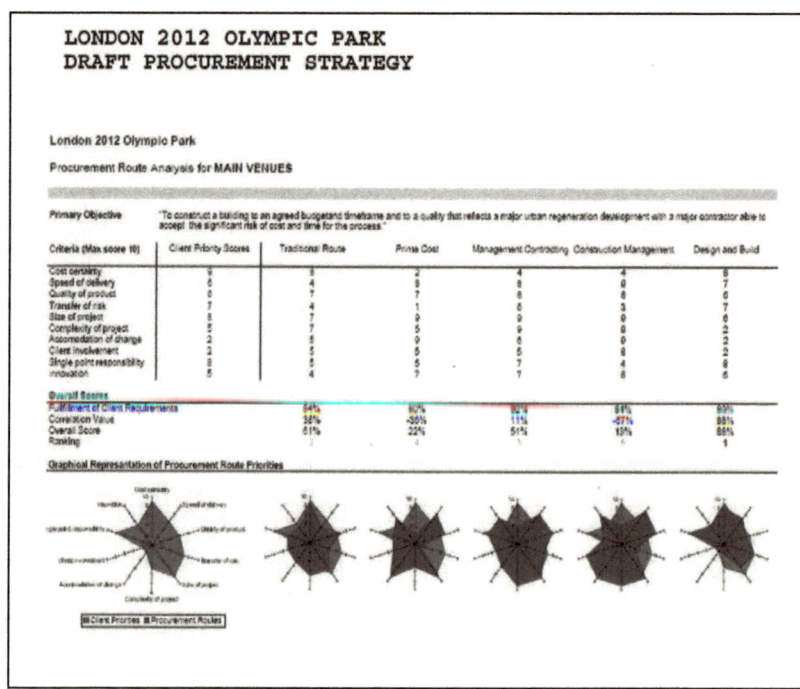

Graphic No 4 – 1: Initial assessment of available procurement options for the main venues in the Olympic Park (ODA 2005)

As can be seen from the above graphic this initial assessment pointed towards Design & Build. Such an assessment was not only carried out for the main venues in the Park but also for the infrastructure works, the statutory authority works, the utilities works, the remediation works, and the demolition and site clearance works.

This initial evaluation was then transformed into a complete procurement strategy taking into consideration not just the more traditional issues of time, cost and quality but requirements from all stakeholders. It is suggested by the respondents that the whole procurement process was very thorough throughout all stages. One interviewee states:

> "Probably the most onerous process in terms of defining...there was an over arching procurement strategy encompassed, transportation, LOCOG issues and the like. Underneath that there was a construction procurement strategy...talking about how we are going to build the park and the venues and then underneath that every project as it is defined, for example the stadium."

The procurement policy of the ODA substantiates this and it is stated that "the most appropriate procedure is to be used in each case, taking into account the legal requirements, specific project characteristics, the likely level of market interest and the timetable requirements" (ODA 2007).

4.2.4 The selected procurement system

The fact that Design & Build was the preferred method for procuring the Olympic Stadium has been shown above. As an important reason for using this method the respondents identified the lack of design information available for the Olympic Stadium at procurement stage and the required early involvement of the contractor for developing the design and construction method.

> "...if you go to the market and say I'd like a full design for an Olympic stadium please you'd struggle to find anyone to design it...there's no such thing really as a trade design for something as sophisticated as a stadium. You also need to get your builder involved early because you need to have that knowledge from him...particularly on an Olympic site where you are going to have lots of other people working around each other and method becomes very important."

It was confirmed that other procurement methods were also considered, such as Construction Management and Traditional Contracting, but these were deemed not appropriate for time reasons.

That the "normal" Design & Build method has its downsides was recognised by the ODA and the responses received from the interviewees draw heavily on the fact that the target cost contract option counteracts these downsides very well.

It was pointed out that by making the decision not to combine this procurement method with a traditional fixed price lump sum contract the ability to make changes was maintained, the monetary premiums associated with the level of risk allocation were removed and the conflict between aesthetic quality and ease of fabrication was resolved.

The above cannot only be applied to the Olympic Stadium but also to the Aquatic Centre, which is also procured under a target cost contract combined with a variant of the "normal" Design & Build. The design for the swimming pool will be novated from the client to the main contractor.

It was expressed that the biggest risk on the Aquatic Centre is, apart from preserving the design quality all the way through to the completion of the facility, the establishment of a good working relationship between the contractor and the designer. The fact that a lot of consideration went into this is highlighted by a quote from one of the participants:

> "I mean one of the major issues for that procurement was bringing together a premier league contractor and putting it with an architect such as Zaha Hadid was how you did that and that whole process was considered at great length."

Turning to the selection of the contractors for both venues it was suggested by the interview participants that there were good reasons for using "Negotiation" on the Olympic Stadium and "Competitive Dialogue" on the Aquatic Centre.

When selecting contractors via "negotiation" the question of value for money arises, but the reasons for why this type of selection was used on the Stadium are best expressed by the participants' quotes below:

> "...limited depth of supply chain in the type of contractor that could deliver it, there are not many people that can do big stadia in this country"

> "I don't actually think they had a choice given the space of the market over the last 2 years, the market has been very over heated, there has been huge amount of construction work going on, particularly in the south-east, particularly in the city,

The Aquatic Centre suffered from a lack of interest and only 3 bidders entered into the "Competitive Dialogue" with the ODA. It was stressed in the literature review that the biggest risk of this process is that not enough bidders are left in the end for there to be a genuine competition. In case of the Aquatic Centre, this is exactly what happened; only one out of the initial three bidders was left to submit a bid for the construction of the swimming pool. However, there is consensus amongst the respondents that this selection method nevertheless was a good choice, which is pointed out be the below quotes:

> *"That sort of dialogue drives out whether you really have got a true partnering culture on board..."*

> *"...at that scale of project I think that level of interview probably works better."*

> *"It is very important that you understand the people on the other side of the fence, how do they work... how do they operate, what motivates them..."*

The final element of the chosen procurement system for the Olympic Stadium and the Aquatic Centre is the target cost form of contract. It has already been mentioned that the target cost contract option is seen to counteract some of the issues encountered with Design & Build and a fixed price lump sum contract. It was suggested by the interviewees that the industry considered these projects as high risk undertakings compared to other projects that were available on the construction market at that time. In addition, the experience from Wembley Stadium, the attached prestige and the challenging design, which for the Stadium was not sufficiently detailed and for the Aquatic Centre came together with a Novation request. This resulted in high premiums for the venues and raising cost.

> *"I think, because the design and build cost were coming in so high because McAlpine put so much contingency in there because... it is very iconic, it is a one off, we have not done this design before...lets put a big contingency in there..."*

> *"...cover the risk of the unknown and the difficulties associated with it and the profile of it...contractors looking at this job would perceive it to be similar as Wembley and want to price the risk accordingly and also put a premium on because it is an opportunity."*

It was expressed that the use of Design & Build together with a target cost contract cannot only cut down the above mentioned premiums but can also make the approach more flexible, it incentivises the contractor, allows an early start on site and paves the way for a less adversarial relationship between the parties.

The respondents' answers have also confirmed that the ODA has realised that the client's relatively great exposure to cost risk is the main downside of this form of contract and that it is important to administer the contract appropriately.

> "I mean again target costs are fine provided you administer the contract robustly… If you are not managing change and valuation of compensation events etc that is where it can run away from you."

> "…but you as client have to make sure that you have somebody on your side to sign all those things off and actually understand what you (as client) are trying to achieve."

The ODA's delivery partner CLM is therefore seen to be best placed to manage this form of contract and most of the respondents conclude that if it is used correctly target cost contracts are a cooperative approach to procurement that can deliver projects on time and to budget.

5 Analysis and Discussion

5.1 Introduction

The successful delivery of these venues will ultimately depend on how well the chosen procurement systems perform against the set targets. This chapter will analyse and discuss the collected data with regard to these targets. Additionally, the main research objectives will be addressed, which will pave the way for conclusions to be drawn in the final stage of this study.

5.2 The client

During the empirical investigation it was established that the ODA is not seen to be a typical public client. It is an organisation that had to be created very quickly and does not have a real history. Moreover, the level of experience within the ODA as to the implementation of large scale projects was not seen to be sufficient at the initial stages. In order to become an effective leader people with the necessary knowledge and understanding had to be brought in. Cherns and Bryant (1984) refer to those structures as temporary multi-organisations, which are groups of highly qualified individualistic professionals who are more or less unknown to each other. Cherns and Bryant highlight that due to the inherent complexity in such organisations the client's own organisational arrangements to administer a construction project can become an important cornerstone for successful delivery. On the "Holyrood" project, for example, this was identified as one of the major problems because the client structure was found to be amorphous and difficult to understand. Based on the collected data, such problems are unlikely to occur on these two projects. Despite the fact that the ODA had to be transformed into a knowledgeable construction client and a lot of the management responsibility has been divested to their delivery partner CLM, the structure of the ODA is understandable and appears to be accompanied by set processes and good channels of communication.

It was however confirmed that one of the downsides of the ODA is its inherent bureaucracy. This is due to the fact that the ODA as a public client has to make sure that transparency and formal requirements are adhered to. On the one hand, this has the potential to delay the progress of the projects and also increase the cost due to long-winded decision making processes. On the other hand, such formal requirements can be seen as an assurance that these projects are administered appropriately and procurement decisions are made with the confidence that all strategic options were identified.

One such formal requirement is that the ODA has to follow the OGC gateway process for all its procurement activities, which includes the Olympic Stadium and the Aquatic Centre. The driver behind this process is the need to provide best value for money and to complete the project without exceeding the available budget. To meet its objective this gateway process requires that every procurement strategy has to be based on a solid business case that demonstrates a clear understanding of the funds required and any budget constraints, taking into consideration the design and purpose of the building and the required completion time. This is an important requirement due to the fact that especially on an Olympic project, where innovation is an integral part of the design and construction process and the client is of public nature, the project does involve higher costs and more varied and complex sources of funds. Where funding involves the taxpayer any budget issues can pose a large risk to projects and their overall outcome in terms of quality and time (CABE 2003).

That such a business case was in place, following this gateway process, for the Olympic Stadium and the Aquatic Centre, , was confirmed during the interviews and is also supported by the ODA's procurement policy. For that reason, the author suggests that such formal requirements and transparency obligations are an assurance that procurement decisions were made based on a clear understanding of what is required in terms of time, cost and quality.

5.3 The selection of the procurement method

It was highlighted by the author in the literature review that a systematic and independent approach is required to ensure that the later selection is not biased and that the project criteria are matched to the characteristics of the most suitable procurement method.

The data collected during the empirical investigation described the selection process of the ODA as very thorough. Before any procurement decision were made for the Stadium and the Aquatic Centre the ODA seemingly applied a very systematic approach.

It was confirmed that the procurement process for both venues not only started very early but was also assisted by an analytical examination of the available procurement options. The ODA assessed and evaluated the available options with regard to their suitability for these projects together with independent consultants, at a point in time where their delivery partner CLM was not even involved. The involvement of such external consultants suggests that the selection of the procurement methods was carried out in a disciplined and objective manner.

Developing an informed strategy, especially for large projects like the Olympic venues, is not an easy task but the ODA appears to have understood its responsibilities and the tasks involved in managing this selection process.

5.4 The selected procurement system

5.4.1 The procurement method

Based on the above described selection process, Design & Build and Novated Design & Build were chosen as the preferred procurement methods to build the centre pieces of the Olympic Park, the Stadium and the Aquatic Centre.

As a main reason for choosing Design & Build for the Olympic Stadium it was expressed that the ODA did not have the benefit of a concept design for this venue and only little design information was available at procurement stage. However, the Stadium is a complex structure considering the requirement to accommodate the planned reduction in size for its legacy usage. Therefore, the collected data emphasises the fact that the ODA required the early involvement of a contractor for developing the design and construction method for this venue.

In the context of the key criteria of time the author feels that this makes sense due to the fact that a programme had to be established that meets the immovable deadline, taking into consideration the timeframes for both the design and the construction.

In a study carried out by Moorledge et.al (1996) it is argued that the time for construction is a direct consequence of the logic of one design solution for any design problem. This, according to Moorledge et al., is due to the fact that for a given design there will be a logical way to build. This logic cannot be changed except by using alternative methods of construction such as off-site prefabrication. Hence, the early involvement of a contractor, who understands the degree of complexity involved in a project like an Olympic Stadium from the very start and who can influence the solution as it develops, is indeed of benefit.

In addition, the Olympic Stadium is a high profile public project and Kumaraswamy and Chan (1995) found that the overall timescales of many such projects appear to be established as a consequence of political considerations, rather than an objective assessment of durations. They suggest that the risk of unachievable project durations under economic, political or commercial pressures can be reduced if the contractor is being consulted early, at a point where the construction and pre-planning times are fixed.

The idea of bringing the contractor in early and to get advice on design and construction logic was not the motivation for using Novated Design & Build on the Aquatic Centre. The collected data confirmed that the development of this building was different and it had been commissioned in terms of its design long before the Olympics were awarded to London. Because of these circumstances the ODA's motivation was to preserve the design quality of this venue throughout all project stages. The author can therefore not resist stating that the decision to novate the design to the contractor made perfect sense.

It is the author's opinion that the design and appearance of the venues play a central part in their purpose, which is to portray design excellence in the country holding the games. It is therefore essential that the client chooses the right procurement method that will ensure that astonishing aesthetics are achieved.

It was argued in the media that the ODA did not put the same focus on the aesthetic quality of the Stadium by using Design & Build. This is owed to the fact that it is assumed that the contractor will do what is necessary to increase the buildability of the project rather than aesthetics under the lump sum contract for reasons of cost. However, the ODA seems to have realised that there can be a conflict between aesthetic quality and ease of fabrication when using this method "off the shelf" and have decided to use it with a target cost contract instead.

In fact, it was confirmed during the empirical investigation that the use of the target cost contract option for both venues has counteracted a lot of the pitfalls that are normally encountered with Design & Build and Novated Design & Build respectively.

In particular the allocation of risk in both methods, when used with a fixed price lump sum contract, is a big concern and attracts huge amounts risk premiums.

This is why at first glance both methods of procurement were seen as very risk adverse approaches and the initial concerns of the media about quality and cost of the venues are understandable. The collected data suggests that the ODA had indeed planned to do what is very typical for a public client, which is to try to transfer most of the risk to the contractor.

Only when initial cost estimates were received for the Olympic Stadium and the Aquatic Centre the ODA seemed to have realised that the transfer of risk on such high profile and complex projects has a significant cost attached to it and that value for money in the finished project can be considerably enhanced through a more responsible allocation of risk.
The results of the empirical investigation show that the target cost contract option is seen to be a good alternative which steps towards a more responsible allocation of risk, especially when used with Design & Build.

Therefore, the author suggests that by using Design & Build with a different form of contract not only enables the client to reduce the risk premiums but also allows an easier implementation of changes at a later stage. These are normally two of the main downsides of Design & Build method of procurement.

5.4.2 The contractor selection

For selecting its contractors the ODA has used the "Negotiation" and the "Competitive Dialogue" route, which are two out of the four available selection methods for a public client under the EU procurement rules.

In terms of finding the right partner both selection methods can certainly be classed as very effective. The collected data shows that these selection methods do have the advantage of giving the client the chance to get to know the people on the other side of the table before a contract is signed. Especially on the Aquatic Centre where the ODA had to establish a good working relationship between the contractor and the designer, because of the novated design, it can certainly be seen as beneficial to go through an interview process.

However, both selection methods are heavily regulated and especially on the use of the "Negotiation" route the EU regulations are very restrictive in terms of justifying its use. This is because under both methods there are concerns about the distortion of competition.

It cannot be said whether or not competition was distorted and in how far other European and international companies were invited and encouraged to submit a bid for the Olympic Stadium and the Aquatic Centre. However, by looking at what happened on both venues it needs to be questioned whether or not it was the right decision to use these methods in terms of cost and achieving value for money.

The Aquatic Centre had three potential bidders involved, but only one company was left when the "Dialogue" was completed. The Stadium, on the other hand, had only one bidder altogether. Putting it briefly, there were not enough bidders left in the end for there to be any competition.

The ultimate objective of the public client is however to obtain maximum value for money. Considering this objective but having such a lack of competition is definitely concerning. The competitiveness of the submitted prices for both venues is therefore in question, which is an important corner stone to ensure that value for money is obtained.

However, it is arguable if the ODA put itself into this situation. External influences such as the very strong construction market at that time and the bad experience at Wembley might have indeed caused a lack of interest in the market place to pursue such high profile projects that undoubtedly have a lot of risk attached to them.

5.4.3 The chosen form of contract

As mentioned earlier in this chapter the target cost form of contract is seen to neutralise a lot of the issues encountered with Design & Build. It was confirmed during the empirical investigation that risk premiums can be reduced, the conflict between aesthetic quality and ease of fabrication can be removed and adversarial relationship, which are encouraged by the traditional fixed price job, can be avoided because the contractor is incentivised to work with the client.

However, for these and other benefits to be realised a fair sharing of risk, an effective gain/pain share mechanism and an appropriate administration of the contract are required.

Having spoken to very senior people during this investigation it was confirmed that it is the intention of the ODA to establish a cooperative approach to all its procurement activities.

The contractor selection via interviews supports this intention because such a selection process puts particular emphasis on the establishment of a good working relationship. Such an approach to procurement lets the author assume that the form of contract is reflecting this intention and risk sharing and gain/pain share incentives are set up in a fair manner.

The data also substantiates the view that an appropriate system is in place to administer the contract for both venues and that the ODA have the right partner on board to manage it. Their delivery partner CLM is able to ensure that compensation events are managed effectively, that the target cost is kept current and that sight of the target is not lost.

However, the author is concerned about the setting of the target cost. As can be seen from the results, the target cost for the Stadium and the Aquatic Centre was developed under the absence of any competition. This might result in the situation where the contractor is not motivated enough to work efficiently because the target cost is set too high and the client pays more than required. On the other hand, should the target cost be too low there is a good chance that the contractor will try to recover costs by compensation events and by adjusting the target cost. In summary, there remains a high degree of uncertainty over the accuracy of the target cost and it is unclear if the ODA was able to modulate their exposure to this cost risk, which is a downside of this form of contract.

6 Summary and Conclusion

It is again worth mentioning that the main motivation for this study was the debate in the media as to whether the procurement decisions made by the ODA can deliver the main venues in the Olympic Park successfully. In particular, the use of the Design & Build procurement method and the form of tender have caused concerns about the final costs and the design quality of the Olympic Stadium and the Aquatic Centre.

It was therefore the ultimate aim of this study to determine whether the chosen procurement strategies are the right choice for delivering both venues in time, on budget and to the required design and build quality.

In order to answer this question it was one of the author's research objectives to obtain a better understanding of construction procurement and the key areas affecting the project success. In particular, the study sought to identify best practice in selecting a procurement method and to make out the pitfalls and the shortcomings of the particular elements within the procurement strategies that are used for the Olympic Stadium and the Aquatic Centre.

This objective was achieved with the literature review, which has successfully identified the key areas for a successful project delivery. This review has provided a research framework and has informed the scope and detail of the empirical investigation.

Another objective of this study was to understand why these procurement decisions were made by the ODA and to identify if best practice was followed during the procurement process and if the pitfalls of the chosen strategies were counteracted. Finally, the author intended to determine if the chosen procurement strategies fit the client and the project.

The above was successfully achieved by conducting semi-structured interviews with experts of the industry and people involved in the procurement activities for the Olympic venues. This qualitative primary data collection method and the subsequent analysis provided invaluable findings so that the following conclusions can be drawn:

- The ODA can be seen as an experienced construction client considering the team around it and its delivery partner CLM. There was a clear understanding of what is required in terms of cost, time and quality for both venues. Clear business cases were established before any decisions were made. The procurement decisions were made under careful consideration of all available options and the ODA has followed best practice and undergone a very thorough process in the selection process for both venues.

- The pitfalls of both procurement methods, Design & Build and Novated Design & Build, were recognised and counteracted by the ODA. This particularly shows the fact that the standard fixed price lump sum contract was not used but a target cost contract.

- The decision to use Design & Build for the Olympic Stadium was well considered and made sense considering the complexity of this structure and the available design information at tender stage, which required the early involvement of the contractor to advise on design construction logic.

- The choice to use Novated Design & Build will preserve the design quality of the Aquatic Centre.

- The contractor selection on the Stadium and the Aquatic Centre needs to be seen in the context of the conditions of the construction market at the time of procuring the Olympic Stadium and the Aquatic Centre. Despite the fact that the lack of competition causes value for money concerns both methods of contractor selection are considered appropriate for such projects.

- The target cost contract is an appropriate form of contract for these two projects. It neutralises a lot of the downsides of Design & Build and Novated Design & Build. Appropriate systems are in place to administer and manage this contract and fair risk sharing and gain/pain share mechanisms can be assumed.

Based on the evidence found the author concludes that the chosen procurement strategies fit the client and the project and are the right choice for delivering the Olympic Stadium and the Aquatic Centre on time and to the required design quality.

With regard to the budget there remains a high degree of uncertainty. Further research is required to ascertain how accurate the target costs are and how these targets were developed. Only then can a clear conclusion be drawn.

However, assuming that the target costs for both venues were set accurately the author cannot resist the impression that the ODA has a very good chance to meet them. This is because the ODA has the means to manage the contract and the cooperative relationship to execute it successfully.

Finally, it can be said that if the ODA get it right, then the procurement strategies and contractual arrangements for the delivery of the infrastructure projects may well establish a scheme for best practice in the global construction industry and for the procurement of large, public infrastructure projects in the UK. Alternatively, it could turn into the most painful example of how not to do it. Therefore, delivering these projects successfully presents a huge challenge for the UK construction industry.

7 Bibliography

Books, reports, proceedings

Abrahams, K. and Farrell, P. (2003) An investigation into the influence of design and build procurement methods on client value for money, University of Wolverhampton, RICS Research

Arthur, S. and Nazroo, S. (2003) 'Designing fieldwork strategies and materials', in J. Ritchie and J. Lewis (eds.) Qualitative Research Practice, London, pp.109-137, Sage Publications Inc.

Auditor General of Scotland (2000) Report for 2000, Scotland, Scottish Parliament Corporate Body, Para 3.20

Bennett, C. (2006) Construction Industry Lessons from London: 2012 Olympic Games and Wembley Stadium, unknown source

Bennett, J. and Grice A. (1990) Procurement Systems for building, Quantity Surveying Techniques: New Directions, BSP Professional Books

Birrell, G.S. (1992) 'Choosing between building procurement approaches', in Concept and Decision Factors in Management, Quality and Economics in Building, Bezelga, A. and Brandon, P. (eds), London, Spon Press

Blake, Adam (2005) The Economic Impact of the London Olympics, Nottingham, Nottingham University Business School

Bowen, P.A. (1993) A communication-based approach to price modelling and price forecasting in the design phase of the traditional building procurement process in South Africa, Ph.D. Thesis, University of Port Elizabeth

Bower, D. (2003) Management of Procurement, London, Thomas Telford Ltd

Bunni, N.G. (1985) The spectrum of risk in construction, Report of the Standing Committee on Professional Liability, Lausanne, Fédération Internationale des Ingénieurs Conseils

CABE (2003) Creating Excellent Buildings: A guide for clients, London, The Commission for Architecture and the Built Environment

Centre for Strategic Studies in Construction (1988) Building Britain 2000, Reading, University of Reading

Construction Round Table (1995), Thinking about Building, London, The Business Round Table

Cox, A. and Townsend, M. (1998) Strategic Procurement in construction, London, Thomas Telford, p 37

Department for Culture Media and Sport [DCMS] (2000) Better Public Buildings: a proud legacy for the future, London, DCMS

De Vaus, D. (2002) Surveys in Social Research, 5th Ed., London, Routledge

Duffy, F. (1992) The changing workplace, *location unknown*, Architectural Design and Technology Press

Eggleston, B. (2006) The NEC3 Engineering and Construction Contract: A Commentary, 2nd Edition, Oxford, Blackwell Publishing

Essex, Stephen and Chalkley, Brian (2006) Olympic Games: catalyst of urban change, Plymouth, Department of Geographical Sciences, University if Plymouth

Esterberg, K.G. (2002) Qualitative methods in social research, Boston, McGraw Hill

Flick, U. (1998) An Introduction to Qualitative Research, Thousand Oaks, CA: Sage Publications Inc.

Flyvbjerg, B. (2005) Policy Planning for large Infrastructure Projects: Problems, Causes, Cures, London, World Bank Policy Research Working Paper 3781

Flyvbjerg, B., Bruzelius N. and Rothengatter W. (2007) Megaprojects and Risk, Cambridge, Cambridge University Press

Franks, J. (1990) Building Procurement Systems, A Guide to Building Project management, Ascot, Chartered Institute of Building

Fraser, P. [Lord] (2004) The Holyrood Inquiry, Scotland, Scottish Parliament Corporate Body

Freshfields Associates (2005) 'Competitive dialogue: the EU's new procurement procedure', London, Freshfields Bruckhaus Deringer

Gidado, K. and Shamsaddin, A. (2004) Suitability of different design and build configurations for procurement of buildings, University of Brighton, RICS Research

Gray, C. and Hughes, W.P. (2000) Building design management, London, Arnold

Hackett, J. (1998) Design and Build. Uses and Abuses, London, LLP Reference Publishing

Handy, C.B. (1985) Understanding Organisations, Harmondsworth, Penguin

Hanif, Tahir (2007) Module: Construction Organisation, Procurement Strategies – Lecture notes and slides, London, Kingston University

Howell, P. and Hardcastle, C. (1995) Integrating and managing the project team, Glasgow Caledonian University, RICS Research

Hewitt, R.A. (1985) The procurement of buildings: proposals to improve the performance of the industry, unpublished project report submitted to the college of Estate Management for the RICS Diploma in Project Management

HM Treasury, Central Unit on Purchasing (1992) Guidance Document No 36: Contract Strategy Selection for Major Projects, London, HMSO

Hibberd, P. and Djerbarni R. (1996) Criteria of choice for Procurement Methods, Mid Glamorgan, University of Glamorgan

Hickson, D.J., Butler, R.J., Cray, D., Mallory, G.R. and Wilson, D.C. (1986) Top decisions: Strategic decision-making in organisations, 1st Edition, San Francisco C.A., Jossey-Bass Publishers

Howard, J., Packer, A. and Tate B. (1995) Determining the influence of design attributes on construction cost, University of Portsmouth, RICS Research

JCT Joint Contracts Tribunal (2008) Practice Note – Deciding on the appropriate JCT contract, London, Sweet & Maxwell ltd

Jericevich, M. (2008) Securing a Sustainable and competitive Business Strategy for Daly International in the UK mobile telecoms network infrastructure industry, Warrington University, MBA dissertation

Kelly, J.R., MacPherson, S. and Male, S. (1992) The Briefing Process: A Review and Critique, London, RICS Research

Kelly, J. and Male, S. (2001) A value management approach to aligning the team to the client's value system, Caledonian University, RICS Research

Latham, M. (1994) Constructing the team: Joint Review of Procurement and Contractual Arrangements in the United Kingdom Construction Industry, London, HMSO

Masterman, J. (1992) An introduction to Building procurement systems, London, Spon Press

Masterman, J. (2002) Building Procurement Systems, 1st Edition, London, Spon Press

Masterman, J. (2004) Building Procurement Systems, 2nd Edition, London, Spon Press

McDermott, P. (1999) 'Strategic and emergent issues in construction procurement', in Procurement Systems: A Guide to Best Practice in Construction, Rowlinson, S.M. and McDermott, P., London, E & FN Spon

Moorledge, R. Bassett D., Sharif, A. (1996) Client expectation and construction industry performance, Nottingham Trent University, RICS Research

Moorledge, R. and Sharif, A. (1996) Strategies for procurement: development of choice and implementation methodology, Nottingham Trent University, RICS Research

Moorledge, R., Smith A. and Kashiwagi D.T. (2006) Building Procurement, 1st Edition, Oxford, Blackwell Publishing

Mosey, D. (1998) Design and Build in action, Oxford, Chandos Publishing

Murdoch, J. and Hughes, W. (2005) Construction Contracts: Law and Management, 3rd Edition, London, Spon Press

Neuman, W.L. (2000) Social Research Methods – Qualitative and Quantitative Approaches, 4th edition, Boston, Allyn and Bacon

Office of Government Commerce [OGC] (2007) Achieving Excellence in Construction: Procurement Guide 06 – Procurement and contract strategies, London, OGC

Office of Government Commerce [OGC] (2007) Achieving Excellence in Construction: Procurement Guide 09 – Design Quality, London, OGC

Olympic Delivery Authority [ODA] (2005) Draft Procurement Strategy, unpublished

Olympic Delivery Authority [ODA] (2007) Procurement Policy, London, ODA

Patton, M. Q. (1990) Qualitative Evaluation and Research Methods, 2nd edition, Newbury Park, CA: Sage Publications Inc.

Ratnasabapathy S. and Rameezdeen R. (2006) 'A multiple decisive factor model for construction procurement system selection', London, RICS research

Rowlinson, Steve (1999) Procurement Systems in Construction: A Guide to Best Practice, London, Routledge

Ryan, E. C. (2005) Course notes – Construction Organisation, London, Kingston University

Siddiqui, I. (1996) Novation: and its comparison with common forms of building procurement, Ascot, CIOB

Smith, N.J. (1999) Managing Risk in Construction Projects, Oxford, Blackwell Science Publishing

Stephenson, R.J. (1996) Project Partnering fir the Design and Construction Industry, Chichester, John Wiley & Sons

Strauss, A. and Corbin, J. (1990) Basics of qualitative research: Grounded theory procedures and techniques. Newbury Park, CA: Sage Publications Inc.

Trochim, W.M.K (2000) The Research Methods Knowledge Base, 2nd edition, Cincinnati, Atomic Dog Publishing

Turner, A. (1996) Building Procurement, 2nd Edition, Hampshire, Macmillan Building Distribution Ltd.

Turner, D. (1995) Design and Build Contract Practice, 2nd Edition, Harlow, Longman

Walker, A. (1996) Project Management in Construction, 3rd Edition, London, Collins

Walker, D and Hampson, K (2002) Procurement Strategies – A relationship based approach, Oxford, Blackwell Publishing

Wardrop, R. (1996) The design of the building procurement process, University of Paisley, RICS Research

Journal Articles

Abrahamson, M. (1984) 'Risk Management, International Construction Law Review', vol 1, No 3, pp 241 - 264

Adam, S. (1999) 'Update on design and build', Architects Journal, 3 June, 46-8

Bennett, J. and Flanagan, R. (1983) 'For the good of the client', Building, no. 27, pp 26 – 27

Black, O. (2006) 'Talking yourself into it', Building, no 41, pp unknown

Chan, A.P.C. et al. (2001) 'Application of Delphi method in selection of procurement systems for construction projects', Construction Management and Economics, Vol 19, pp 699 – 718

Chang, C.Y. and Ive, G. (2002) 'Rethinking the Multi Attribute Utility Approach based procurement selection technique', Construction Management and Economics, Vol 20, pp 275 – 437

Cherns, A.B. and Bryant, D. T. (1984) 'Studying the client's role in construction management', Construction Management and Economics, Vol 2, No 2, pp 177 – 184

Eaglesham, Jean (2006) Olympics row deepens over Pounds 400m bill for cost control, Financial Times, Nov 22, p.1

Flanagan, R. (1981) 'Change the system, Building', dated: 20th March

Gillespie, B. (1994) 'Procurement Route', Building, vol 46, page unknown

Glancy, Lisa (2008) Olympics effect fuels work boom as clients rush to get jobs finished, Construction News, February 2008, issue 7057, p.13

Hyett, S. (1996) 'Let's look at some alternatives to CCT', The Architects Journal, Vol. unknown

Hibberd, P., Jagger, D. and Morledge, R. (1995) Editorial Comment, Journal of Construction Procurement

Hughes, WP (1992) 'An analysis of design and build contracts', Construction Information File No.6, Chartered Institute of Building, Ascot

Kumaraswamy, M. M. and Chan, W.M. (1995) 'Determinants of Construction Duration', Construction Management and Economics, Vol. 13, pp 209 – 217

Love, P.E.D., Skitmore, M. and Earl, G. (1998) 'Selecting a suitable procurement method for a building project', Construction Management and Economics, Vol. 16, p. unknown

Luu, S.D.T. and Chen, S. (2003) 'Parameters governing the selection of procurement system', Construction and Architectural Management, vol 10 (3), pp 209 - 218

Mintzberg, H., Raisinghani, D. and Theoret, A. (1976) 'The structure of unstructured decision making processes', Administrative Science Quarterly, vol 1, p. 246

Mohsini, R. and Davidson, C.H. (1989) 'Building procurement = key to improved performance', paper to Chartered Institute of Building, International Workshop on Contractual Procedures, Liverpool

Moorledge, R. (1987) 'The effective choice of building procurement method', Chartered Surveyor, July, p 26.

Naphiet, H. and Naphiet, J. (1985) 'A comparison of contractual arrangements for building projects', Construction Management and Economics, Vol. 3, pp 217-231

Nutt, P.C. (1984) 'Types of organisational decision processes', Administrative Science Quarterly, vol 29, p.414

Rawlinson, S. (2007) 'Procurement – Target price contracts', Building, issue 37

Rogers, David (2008) Huge Logistics exercise will leave a lasting legacy, Construction News, February 2008, issue 7057, p.13

Skitmore, R.M. and Marsden, D.E. (1988) 'Which procurement system? Towards a universal procurement selection technique', Construction Management and Economics, vol. 6, pp 71 – 89

Smit, J. (1995) Projecting success, New Builder, 17th March

Sherwood, Bob (2006) Architect 'wrong' to attack Olympics procurement process, Financial Times, Nov 10, p. unknown

Walker, D.T.H. (1995) 'The influence of client and project team relationships upon construction procurement performance', Journal of Construction Procurement, no 1, pp 42-55

WWW Documents

Ashgate (2008) [www] Available at
www.ashgate.com/subject_area/downloads/Sample_Chapters/projmgt8_ch1.pdf

Building (2006) RIBA president warns of 'tarmac and plasterboard' Olympics [www] Available at http://www.building.co.uk/story.asp?storycode=3078298 (Accessed 24th February 2007)

Building (2006) Olympic Media Centre to be design-and-build contract [www] Available at http://www.building.co.uk/story.asp?storycode=3076617 (Accessed 24th February 2007)

Building (2008) New 2012 Olympic stadium images [www] Available at http://www.bdonline.co.uk/story.asp?storycode=3121707&origin=BDbreakingnews (Accessed 2th September 2008)

Construction Excellence (2004) Procurement [www] Available at http://www.constructingexcellence.org.uk/procurement (Accessed 24th February 2007)

Matthews, G. (2006) The 2012 Olympics - Time to pick up the pace? [www] Available at http://www.qsweek.com/nav?page=qsweek.contentspage&fixture_page=5087324&resource=5087324&view_resource=5087324 (Accessed 1st April 2007)

NCE (2007) Auditors orders ODA to open up on programme plans [www] Available at http://www.nce.co.uk/structures/news/auditor_orders_oda_to_open_up_on_programme_plans.html (Accessed 25th July 2007)

NCE (2008) London 2012 budget an lack of legacy plan attacked [www] Available at http://www.nce.co.uk/printPage.html?pageid=1219682 (Accessed 22nd April 2008)

Wikipedia (2008) Construction [www] Available at
http://en.wikipedia.org/wiki/Construction#Procurement
(Accessed 8th February 2008)

University of Nottingham (2008) [www] Available at
http://ibis.nott.ac.uk/guidelines/ch8/chap8-4.html
(Accessed 31st August 2008)

8 Appendicies

Appendix 1 - Interview Schedule

Part - 1 Question No.	Question Content	
1	How would you describe the characteristics and culture of the ODA as an organisation?	
	Guidance Points for discussion:	- client structure - Processes - channels of communications - project environment, i.e. the "landscape" of the project - clear responsibilities
2	How would you rate the ODA's internal experience and knowledge in procuring mega-projects?	
	Guidance Points for discussion:	- the people that are working for the ODA
3	How do you think the ODA have carried out the briefing process to their procurement team for the 2 venues at the initial stages of the procurement process?	
	Guidance Points for discussion:	- was there a comprehensive business case for each of the 2 venues - was there and advisory firm that helped the ODA to clarify and prioritise their objectives
4	How would you describe the ODA's understanding or appreciation of the project in terms of the key criteria of time, cost and quality before the procurement decisions were made?	
	Guidance Points for discussion:	- the funds required to build the venues - appreciation of complexity - the risk of an evolving and changing project scope over time - the time durations required to design & build the projects - the design quality required in terms of aesthetics
5	Do you believe that there is & was a thorough understanding of the importance to decide on the relative significance of cost, time and quality within the ODA? What do you regard as the ODA's 1st, 2nd, and 3rd priority?	
	Guidance Points for discussion:	- What is the ODA most likely to sacrifice to meet the immovable deadline for the completion of the venues – Design quality or cost?
6	How would you describe the ODA's overall attitude towards risk and risk allocation?	
	Guidance Points for discussion:	- Was there a detailed identification of all risks before procurement decision were made - Allocation strategy, i.e. transfer or retention - value for money

7	Are you aware of how the selection of the procurement methods was carried out by the ODA?	

	Guidance Points for discussion:	- Timing of the procurement process – was it carried out early enough
		- What Method was used to evaluate the options:
		    o Analytical evaluation,
		    o theoretical models
		    o historical ("it has worked on other projects")
		- What role did policy compliance play
		- Were all available options considered and properly discounted
		- Are there written records of this process

8	Do you believe that the Design & Build procurement route was the best option for the Olympic Stadium & the Aquatic Centre considering all the other available options?	
	Guidance Points for discussion:	- Construction Management - Management Contracting - Traditional - PFI
9	From your experience what are the potential pitfalls of the Design and build procurement route (and Novated D&B) for such a large project? Do you feel that the ODA is well aware of them and that those pitfalls are dealt with effectively?	
	Guidance Points for discussion:	- Briefing - Evaluation of variations - Risk allocation - Rushed design decisions - Quality of design
10	In your view, why has the ODA chosen to select their contractors via negotiations with hardly any competition and what effects might this have? What selection method would you have advised to the ODA, and why?	
	Guidance Points for discussion:	- Shortlisting of contractors - Target cost contract - Distortion of competition - Final Cost - Value for money
11	Would you say that a target cost contract is the right contractual "back up" for a Design & Build project? Where do you see potential problems?	
	Guidance Points for discussion:	- Normally fixed price - Complexity of projects - Level of project definition at tender stage - Setting of realistic target price - No competition at tender stage
12	How would you assess the "effectiveness" of the target cost contract used for the venues?	
	Guidance Points for discussion:	- Risk allocation - Gain\pain share arrangements - Commitment to target price - incentives

Appendix 2 – Interview Invitation Letter

[Name & Address of interviewee]   **Dirk von Plessen**

[Date]

INFORMED CONSENT FOR PARTICIPATION IN AN MSC RESEARCH PROJECT

Dear [Name of interviewee],

This letter serves to confirm an invitation to consider participating in a study I am conducting as part of my Master's Degree in 'Management in Construction' at Kingston University London under the supervision of my tutor Alan Ellis. I would like to take this opportunity to provide you with more information about this project and what your involvement would entail if you decide to take part.

The International Olympic Committee announced on the 6th July 2005 that the Games of the 30th Olympiad in 2012 go to the city of London. New sporting facilities are already under construction to host this prestigious event; amongst them are the two flagship venues the Olympic Stadium and the Aquatic Centre.
The rather difficult task the Olympic Delivery Authority (ODA) is facing is to deliver these two "high-profile" facilities to an immovable deadline, to stay within budget, and at the same time to deliver the venues with astonishing design and build quality. These are the main criteria against which the success of these projects will be measured.

With this in mind, the ODA have decided to procure the Olympic Stadium under the Design & Build procurement route and it is believed that the Aquatic Centre is procured under Novated Design & Build. In addition, for both venues the NEC3 target cost contract was used and the contractors that are to carry out the works were selected by means of negotiations.
Due to those procurement decisions and the ongoing debate about the rising budget requirements for these projects the critics on the ODA's procurement strategy become louder and it is argued whether or not cost, time and quality targets can be met for the two flagship venues.

The purpose of this study, therefore, is to determine whether the procurement strategies chosen by the ODA are the right choice for delivering the two main venues in the Olympic Park in time, to budget and to the required design and build quality.
This study has the following research objectives:

- Examine if the procurement strategies chosen by the ODA fit the client & the project

- Identify best practice of how a procurement strategy should be selected and examine if this was followed by the ODA whilst making its procurement decisions.

- Identify the pitfalls & weaknesses of the particular procurement strategies used for the Olympic Stadium & the Aquatic Centre and examination if they have been recognised and counteracted by the ODA.

- Identification of the known effects and consequences on cost, time, and quality of the chosen strategies.

The author feels that this study will help to increase the understanding of the procurement decisions made by the ODA.

However, for this research project to be of value some in-depth information of construction procurement and opinions on how the ODA is dealing with this difficult task are vital and cannot be gained through any other method than personal dialogue with experts of the industry that have a good understanding of this matter.

It should be noted that participation in this study is voluntary. It will involve an interview of **approximately 45 minutes** in length to take place in a mutually agreed venue. You may choose to decline to answer any of the interview questions if you wish. Further, you may decide to withdraw from this study at any time, without prejudice, by advising me of your intention to do so. With your permission, the interview will be tape-recorded to facilitate an accurate recording of the responses.

All information you provide is considered completely confidential. Your name will not appear in my dissertation resulting from this study however, with your permission anonymous quotations may be used. There are no known or anticipated risks to you as a participant in this study.

If you have any questions regarding the above, or would like any additional information to assist you in reaching a decision about participation, please contact me or my supervisor on the numbers below:

I would like to assure you that this study meets the requirements of the University's ethical Guidance and Procedures for Undertaking Research Involving Human Subjects.

I very much look forward to speaking with you and I would like to thank you in advance for your assistance in this project.

Kind regards,

Dirk von Plessen
Kingston University London

WRITTEN CONSENT TO PARTICIPATE IN A RESEARCH STUDY

Statement by participant

- I confirm that I have read and understood the information presented in the invitation letter for this interview. I have been informed of the purpose, risks, and benefits of taking part.

- I understand what my involvement will entail and any questions have been answered to my satisfaction.

- I am aware that the interview will be tape recorded to ensure an accurate recording of my responses.

- I understand that my participation is entirely voluntary, and that I can withdraw at any time without prejudice.

- I understand that all information obtained will be confidential.

- I agree that research data gathered for the study may be included in the final dissertation to come from this research, with the understanding that data will be anonymous and that I cannot be identified as a subject.

- Contact information has been provided should I wish to seek further information from the investigator at any time for purposes of clarification.

Participant's Signature:_____

Date: _____/_____/_____

Statement by investigator

- I have explained this project and the implications of participation in it to this participant without bias and I believe that the consent is informed and that he/she understands the implications of participation.

Name of investigator: _____Dirk von Plessen____

Signature of investigator: _____

Date: _____/_____/_____

Appendix 3 – Quotation references - The Project

APPENDIX 3

Source	Quotations / key phrases
4.2.1 The Project	
Type of project	
BOVIS LEND LEASE – RESPONDENT 3	"You have to remember that what the ODA are doing is something in the country that has never been done before"
Ferrovial	"It is a very political project or projects, they have to work under a lot of pressure…"
ODA	"We're buying two and a half thousand contracts" "We've got two hundred and fifty major construction contracts to buy" "So that's the scale of it."
ODA	"The other side of the coin is that the swimming pool is very complex technically, so is an eighty thousand seater stadium"
ODA	"They are one of the top five as we call them and they are the two signature buildings"
BOVIS LEND LEASE – RESPONDENT 2	"…there is over a 100 stakeholders, you got all the different local councils you got the different government departments', you got this person you got hat person, everyone wants there 2 pounds * and hence time expands and they say, we want this, we want that, we want this, and then the price just spirals, spirals and spirals"

ODA	"It's a bit like any big public project you start off with. You start off, you ask for a mini car and someone says they are going to cost you fourteen thousand pounds. By the time everybody in the committee has decided what they want on it it ends up looking very much like Bentley GT and you've ended up paying a hundred and twenty thousand pounds for what you thought was called a mini but actually it's really a Bentley. And that's really what you end up with public procurement more generally."
ODA	"Every time we kick off a procurement …we have actually got to consider the impact on the local community. It includes the diversity of what we are doing, the sustainability of what we are doing in terms of environment, economic particularly in social impacts, so the issues we have to consider are wide and varied."
Requirements in terms of time, cost and quality	
ODA	"Time comes ahead of everything else."
Ferrovial	"…it will be more embarrassing not to have the facility…"
BOVIS LEND LEASE – RESPONDENT 3	"…you cannot have the Olympics two weeks late."
ODA	"We have a huge amount of interest on this. We have to be transparent in every way …because this is all public money we're spending"
BOVIS LEND LEASE – RESPONDENT 3	"…but you look at people's perception of Sydney and the general perception is that it was fantastic, everything was fantastic, you look at people's perception of Greece and it was, well, they got there. So, the prime minister want to be in a place where everybody is going, that was great."

ODA	"I mean at the end of the day the public purse is going to have to pay for it so we can't be wanton with our expenditure." "But we also have to build something that makes a statement about London and about the UK."
BOVIS LEND LEASE – RESPONDENT 3	"…it is a project that is important to the nation and to the government and to the prestige of the government and the nation * and you don't want to give the impression to the rest of the world * that it was on time but not very good "

Appendix 4 – Quotation references - The Client

APPENDIX 4

Source	Quotations / key phrases
4.2.2 The Client	
Type of client & experience	
ODA	*"The ODA is an organisation formed over a very short period of time." / "the ODA themselves have been formed very quickly"*
BOVIS LEND LEASE – RESPONDENT 2	*"I don't think they are a particularly enlightened client…the likes of land securities or even rbs are more what, and baa, are what I call an enlightened client in that they have the technical knowledge within their own organisation relating to the construction of buildings and the construction industry in general."*
BOVIS LEND LEASE – RESPONDENT 2	*"I think because the ODA was an uneducated client, hence they got people like Mace, Deloyds were also involved, and Davis Langdon"*
BAA	*"No you could tell they were a bit like oh! Shit what do we do here and you could tell from the questions they were asking that they are not people that had perhaps been on some major projects. I would have expected perhaps that they have from a governance perspective they would have probably spent a lot more time with some people that delivered some really big, major projects."*

BAA	*"They are an educated client now, the last year but certainly when we were having some initial meetings and it was two years ago it was a bit like they weren't quite sure where they were going sort of thing." / "At that point in time probably whereas now you get the impression, yes they have made that step change."*
ODA	*"As an organisation, well they are and I absolutely support, they are a very effective group of very senior professional people within the construction industry."*
ODA	*"…I think genuinely, Her Majesty's Government has gone out there and bought the best people they could find and are available within construction. They are not there for the money, they are there for being seen to do the best job possible."*
BOVIS LEND LEASE – RESPONDENT 3	*"David Higgins, as being one of the chair of the ODA, as an ex-chairman of lend lease has a huge amount of experience in terms of doing all types construction and property deals."*
ODA	*"The ODA is a builder because the ODA's remit is to construct the Olympics, the theatre whereby we are going to hold the show. They don't run the show, they don't employ the actors or do the scenery, LOCOG does that, ODA builds the theatre."*
ODA	*"The ODA is that they set out to be a thin client and therefore they are a cost effective client in many respects. The ODA is a construction client." / "It is an informed client."*

ODA	"The ODA executive is made up of a raft of senior professionals who have all worked in construction. And so they all have the technical knowledge. They don't need to have the detailed technical knowledge." / "Harold Shipley, for example, the head of construction here, he has run some of the biggest construction organisations in the country. You get a feel that he probably knows his apples when it comes to building things."
Ferrovial	"The ODA as an organisation seems to be quite a high powered organisation, you have lots of experienced and seasoned people in the structure there"/ "it appears that they have, the top guns working on it and I'm sure that they will deliver it, I mean that's the perception" / "as a professional, do they have the internal experience, yes, I think they do" / "So, yeah, I think generally they have got all the right people on board."
BOVIS LEND LEASE – RESPONDENT 1	"...they have divested that whole responsibility to the collaboration with CLM...CLM is their delivery arm whereas the ODA, or else I am wrong, is very much an administrative bureaucratically public sector operation."
BOVIS LEND LEASE – RESPONDENT 2	"CLM...because they are the internal consultant, programme manager that is being used to manage procurement"
ODA	"CLM were appointed on behalf of the ODA to manage and effectively deliver these contracts for them"
BOVIS LEND LEASE – RESPONDENT 2	"...they [ODA] are the pay master, they are actually managing the money on behalf of the government"
BOVIS LEND LEASE – RESPONDENT 3	"....they[ODA] are quite bureaucratic, they are a public organisation, they have a certain responsibility in terms of audit and process, this is kind the nature of the beast we are dealing with, they have an awful lot of advisers, they have an

	awful lot of experts, they have got an awful lot of processes" / "and the number of stakeholders, absolutely * they (ODA) are kind of forced into that place where making dynamic decision doesn't really come into it"
BAA	"You have got a government organisation which seems to have loads of committees involved and people like Transport for London, the local council etc. You have got government as well. You have got a minister allocated to it and yes it's an important event the Olympics but again it just seems to be a lot of governance around what is going to happen and decision by committee if you follow me. So my perception of the ODA and all that sort of area it will be very difficult to make quick decisions. "
ODA	"It is hard to get things agreed but not because of the ODA's make up at all, it is because the ODA has something like eleven organisations auditing it at any one time.
ODA	"...this is all public money we're spending, with a few exceptions, and so everyone has got a vested interest in the Olympic games so naturally you're going to have quite an autocratic approvals process." / "Many decisions on that approval process aren't ODA people though. They are the ODA's masters and other vested stakeholders with a vested interest."

Project Briefing & understanding, prioritisation of need

BOVIS LEND LEASE – RESPONDENT 2	"I would guess pre-London getting it (the Games) that MACE, Davis Langdon, Deloyds, were in there giving advise, offering advise, they obviously had some deliverables, that (identifying the objectives) was one of them amongst some other stuff, so I think there initial advice came from there and that may have influenced their (ODA) thinking, because why wouldn't they believe MACE, Davis Langdon, Deloyds, they are the experts in their industry, they (ODA) probably took that and they obviously discussed it and debated it"
BOVIS LEND LEASE – RESPONDENT 3	"I almost guarantee there would have been a very clear briefing there would have been a very clear business case and they (ODA) would have clearly got it signed off because what they have done is slightly unusual and in order to get the necessary approvals from government and whatever internal limits of authority they have got they would have had to had all that paper in place you know, irrespective of what answer they have got to they would have had to have that sort of stuff in place"
ODA	" … it's a Prince type system they use here and the have project initiation documents driven by business cases as you say." / "So there is a business case in place for the stadium and to be re-affirmed during the construction process. During the stage gate process the ODA uses they use the OGC gateway process for all their projects, okay, and the stadium has been subject to that"
Ferrovial	"…from my perceptions…there has been a lot of help and there has been a lot of consultants around providing advice to these type of organisations "
BOVIS LEND LEASE – RESPONDENT 2	"…they probably had a much better appreciation after having the some of the advisory firms involved"

ODA	*"Because they are one of the top five as we call them and they are the two signature buildings…there was a huge amount of consideration that went into it. The teams here (at the ODA) are very big and for good reason because prior to kicking off a procurement undertaking you need to understand the issues of time, cost and quality at the simplest level."*
BOVIS LEND LEASE – RESPONDENT 3	*"I imagine they would understand them really well"*
BOVIS LEND LEASE – RESPONDENT 3	*"I would suggest, without a doubt, their main criteria is time because that is the only thing that cannot move, everything else can give a little, to a degree, the only that cannot change is time and again I would be staggered if they ODA did not have a full appreciation of all of those issues prior to them entering into anything"*
BOVIS LEND LEASE – RESPONDENT 3	*"cost, that would be the one that goes because time you cannot have the Olympics two weeks late and then for the world to walk around the Olympic Stadium and then go it's very nice but its a bit shabby, isn't it? They just could not allow that to happen, as well, it would be really difficult but they would let cost go and that kind of is the nature of major, major projects" / "it is getting the balance right in terms of the three key elements of time, cost and quality…but if one of them is going to drop out, I imagine it will be cost"*
ODA	*"in terms of quality "Yes, I think the issue is going to be that there is only so much you can do in terms of value engineering and the thing had been value engineered to death." So I think it's going to be T Q C, would be my view."*
Ferrovial	*"So the thing that you would preserve is the deliverability of the facility."*

Understanding of the risks involved and the attitude towards risk allocation	
BOVIS LEND LEASE – RESPONDENT 3	"…with very large infrastructure projects, whereby, essentially it is never been done before, so to try to get a lump sum price from a contractor when the design does not exits or it exist to such an extent that it is meaningless, either means, that the contractor is going to charge an enormous amount of premium for taking the risk or he is going to leave you with the risk and is not going to take it at all…."
BOVIS LEND LEASE – RESPONDENT 1	"…you got to think about it in that it's in the shadow of Wembley, there were so many issues with Wembley with managing the project, a lot of contractors said why on earth would we divert time and resources and risk into a project like that when there is much other valuable work going on out there"
BOVIS LEND LEASE – RESPONDENT 2	"Public sector is generally very risk averse, they generally won't take the risk on but I think because they had a professional team around them they perhaps think that they could manage some of that risk, hence reduce the overall cost"
BOVIS LEND LEASE – RESPONDENT 1	"Because of the lack of information, that is probably what drove them in the first instance, the lack of design information meant that they wanted to go D&B and divest that risk to the Tier 1 contractor."
BOVIS LEND LEASE – RESPONDENT 2	"I think, because the design and build cost were coming in so high because McAlpine put so much contingency in there because, hey, I can see the pretty picture from the outside and that looks like it is going to be a lot of money and because it is very iconic, it is a one off, we have not done this design before, even though we done the Emirate Stadium, lets put a big contingency in there, and they (ODA) have probably gone puhhhhh * and CLM have said, as a professional team, lets

	us take on some of that responsibility so you actually changed the type of contract into a target cost * and we are actually now trying reduce that contingency because we are going to manage it."
BAA	"Yes I think it is probably because the ODA thought shit, we would actually need to convey to the public that we are being a more responsible client and we are managing the risk better, rather than just passing all that risk to the contractor."
BOVIS LEND LEASE – RESPONDENT 3	"I think they have deliberately tried to appear to be fair in what they are trying to do"
BOVIS LEND LEASE – RESPONDENT 3	"…Design & Build is very risk averse, target cost is sort of holding hands again"
ODA	"…the intention isn't to throw all the risks to the contractor"
BOVIS LEND LEASE – RESPONDENT 3	"I get the impression, like many clients they (ODA) want to be in the place where they allocate risk at the right time in the appropriate manner"
ODA	"The Aquatics was a project that actually had a long burn time prior to any procurement activity being undertaken. The London Aquatic Centre is a piece of art work and it stands on its own in terms of business case rather than being an Olympic project."
ODA	"It's got a huge amount of history on it. Bearing in mind that Zaha Hadid was appointed as a designer, such an architect of some repute in many corners, but also obviously there is a lot of controversy about her work and that has meant that there has been an extremely thorough briefing process prior to initiating the procurement of the works…."

| ODA | *"I mean one of the major issues for that procurement was bringing together a premier league contractor and putting it with an architect such as Zaha Hadid was how you did that and that whole process was considered at great length. Any more and I would have said that the public pocket probably wouldn't have wanted to bear it."* |

Appendix 5 – Quotation references - The Method Selection

APPENDIX 5

4.2.3 The Method Selection

The procurement process & the evaluation of procurement options

Source	Quotations / key phrases
BOVIS LEND LEASE – RESPONDENT 2	"…a draft procurement strategy, developed together with an experienced consultancy who have been involved in similar projects before (MACE)" MACE were involved as a consultant more from a Master planning point of view….This strategy was led by one of the leading directors of MACE who is now also working within CLM on the planning front."
BOVIS LEND LEASE – RESPONDENT 2	"This procurement strategy document is a typical assessment of the various types of methods of procurement weather it be Traditional route, prime cost, Management contracting, Construction management, Design & Build; we have some spider diagrams; and we actually established, in a serious of workshops, what the clients primary objectives were in terms of cost, time, health & safety, legacy, etc. etc. and what form of contract is most appropriate. This was split down into main venues, infrastructure works, statutory authority works, utilities, remediation, and also demolition and site clearance. So, I suppose the main venues is the key one, and obviously this points towards D&B." / "but this then (this assessment) suggests they are looking at a design and build type of procurement."
BOVIS LEND LEASE – RESPONDENT 2	"…they followed the OGC guideline, which is what a lot of MACE processes are based on and there is a picture or model of that in there"
ODA	"But it's not just the more traditional issues of time, cost and quality; we've got a score card, a balance score card we use to evaluate all requirements from all our stakeholders."

ODA	*"Probably the most onerous process in terms of defining. What we did was we wrote a procurement policy that set out what we were going to do. Then there was an over arching procurement strategy encompassed, transportation, LOCOG issues and the like. Underneath that there was a construction, an engineering procurement strategy; construction procurement strategy. They're talking about how we are going to build the park and the venues and then underneath that every project as it is defined for example, the stadium."*
ODA	*"...the initial strategy goes through the OGC process and is audited by OGC and then you've got a detailed strategy. These strategies are a hundred percent pages of detailed consideration of all the different aspects, and bear in mind there is only five ways to buy anything in construction"*
ODA	*"As my day job and then set the ball rolling in terms of what the procurement strategies to the Olympics look like. The first ones were the aquatics and construction of bridges and highways and utilities strategies and they were to look the same way and they were compliant with OGC best practice guidance so they all follow the same format. We had a thing called COG, Compliance and Oversight Group, that was basically the Government's sort of, put a number of key professionals in place to make sure of the process"* / *"It was a very thorough process."*

Appendix 6 – Quotation references - The Selected Procurement System

APPENDIX 6

Source	Quotations / key phrases
4.2.4 The Selected System	
Procurement methods - Design & Build and Novated Design & Build	
Why D&B on Olympic Stadium	
BOVIS LEND LEASE – RESPONDENT 1	"Because of the lack of information, that is probably what drove them in the first instance, the lack of design information meant that they wanted to go D&B and divest that risk to the Tier 1 contractor."
ODA	"If you give the text book answer to that, I mean if I go to a university and talk about design and build, you'd say 'oh you get functional buildings don't you' and 'build to a price' and all that sort of stuff. It doesn't really work like that in reality because you can't, if you go to the market and say I'd like a full design for an Olympic stadium please you'd struggle to find anyone to design it." / "HOK Sport would do the concept but someone else would have to do the production design and even if you got full design you still go out to production design for all these large elements because the designers don't hold that capability so there's no such thing really, arguably, as a trade design for something as sophisticated as a stadium. You need also to get your builder involved early because you need to have that knowledge from him."
ODA	"...particularly on an Olympic site where you are going to have lots of other people working around each other and method becomes very important."

ODA	"Also, you need to lock him into his thinking, because it is no good getting a contractor in, using him as a consultant and then letting him go, giving his ideas to somebody else to go and build out because they won't be happy with that."
Ferrovial	"Well a certain degree of control, if you go back to the first one where you have the normal design and build as long as the ODA can specify what they want in terms of their employer's requirements lets say it's whatever, 100-seat stadium for example, there's no roof in it etc., etc. and then the contractor takes those requirements and builds to that requirement then everything is okay really and a contractor who is used to building stadiums can actually use the processes and the procedures and the methods of working which are best suited to that type of work then actually it is a win win situation for both the client and the contractor"
	D&B and other methods
BOVIS LEND LEASE – RESPONDENT 2	"I had the same thoughts * I mean T5, that was cost plus * but it was CM, but not a proper CM * but yeah the risk sat with the client * you had a professional team around to manage that as well as having his own professional team * so, yeah I would say that because they are so unique, so individual * I would have thought from a time perspective, certainly, and, although you got no guarantee of final cost until you get the final cost, until you get final billing * you don't know what the cost is with CM * I think CM might be a more appropriate * but obviously if they are more risk averse * from D&B to CM would be a big jump * in terms of thinking *"
ODA	"On a programme of this scale, you wouldn't necessarily have the ODA going out and buying trade packages." "I'll tell you why, because it takes us nine months to buy anything because you've got EU regulations to go through for most of it because most of it's over three million quid"

ODA	*"What other options have you got, traditional? Traditional, big issue about traditional is that you've got procurement on the critical path. The design stops happening while you are going through the procurement process. These things are in procurement for nine months to a year. You can't stay in the process, we've got 2012 to deliver we haven't got twenty thirteen."*
	D&B pitfalls
ODA	*"The advantages are quite significant when you come to looking at change. What you've got is, you've got a contractor who is incentivised to actually work with you, to continue working and actually delivering what your requirements are in real time, rather than not do work because he hasn't had it approved under variation instruction."*
ODA	*"There's always the risk of back end change but bear in mind that we've not gone out for fixed price lump sum D and B, we have mitigated that to a degree and actually the contract we have chosen to use, the NEC 3 form of contract, is very important in that actually, if used properly it can mitigate issues of time as well as cost"*
ODA	*"…only if it's a fixed D and B change is problem but of course these aren't fixed price D and Bs for that very reason."*
BOVIS LEND LEASE – RESPONDENT 3	*"with very large infrastructure projects, whereby, essentially it is never been done before, so to try to get a lump sum price from a contractor when the design does not exits or it exist to such an extent that it is meaningless, either means, that the contractor is going to charge an enormous amount of premium for taking the risk or he is going to leave you with the risk and is not going to take it at all, and just price what is on drawing, which is next to nothing, you then end up in a situation*

	where you constantly battling, whereas, you can go with target cost and you got to have a set of rules, you stand half a chance to get to a sensible answer and not getting ripped off."
ODA	*"So HOK are not really going to let you build it out of plastic, margarine tins if they are associated with it. So that just simply isn't going to happen because you are assuming that the contractor, his vested interest is reducing quality to increase costs and therefore profitability, but he cannot really do that so well on a target price arrangement. He's actually incentive is to delivery what the client wants, so you've kind of removed a lot of that, sort of like he'll throw up a shed very quickly, very efficiently kind of using D and B."*
	Novated D&B on Aquatic Centre
ODA	*"I mean one of the major issues for that procurement was bringing together a premier league contractor and putting it with an architect such as Zaha Hadid was how you did that and that whole process was considered at great length. Any more and I would have said that the public pocket probably wouldn't have wanted to bear it."*
ODA	*"So, I mean if you play back the kind of D and B functionality, good for sheds bad for iconic buildings like an aquatics kind of theory, what you don't take on board is the fact that you've got a signature architect." / "…and the architect had been appointed for some considerable length of time, had been designing through and she was up to stage D when Balfour Beatty arrived, and the concept was well developed. But you would not be seeing the whole picture because the issue is that you are not going to go and buy a Zaha Hadid building and not let Zaha Hadid influence the functionality and the quality of the outlook. This is going to be a statement this building; it's a piece of artwork."*

Ferrovial	"The Aquatic Centre.....the whole thing rests on the aesthetic appeal of this building, now if you give this to any other contractor and say design and build it might not look like what it does at the moment so to preserve the architectural philosophy and the aesthetics and all that, it is very important that her design is actually novated over the to contractor, so that the contractor then takes that design and builds something that looks like it ..."
Ferrovial	"On the Aquatic Centre to maintain the aesthetics are very important to go novated because I think that is the only way to secure that."
BAA	"So in respects I would see that [Novated Design & Build] perhaps is a better route because I would feel that I as a client would have had more control over the design in the early stages so that in some respects I would hope that because those consultants were working for me that we have got to a point where we were happy and we understood and basically when you are procuring your main contractor they know that they have to pick up those consultants but in some respects I would have hoped that because that design to have got to a reasonably mature stage that you have also have a little bit more certainty around cost and what you are actually getting for your money so therefore I would hope that that project is more likely to come in on time and on budget and in the way what you wanted basically."
ODA	"I think there is an issue about design and build with Novation when you kind of get given what you are given and there is obviously a fear within the contractor's side of the relationship that basically he's going to be hauled by nose by the senior architect to deliver something that is totally unaffordable for him and his client and it's an issue about how he controls the architect."
BAA	"Can the design supply chain work with the contractor? So there is mainly a big relationship ticket."

Contractor Selection: Negotiation & Competitive Dialogue

Negotiation on the Olympic Stadium

BOVIS LEND LEASE – RESPONDENT 1	"…even those big boys that got the capacity to do this job, if they did this it would exclude them from doing any other works, so there is even more risk involved, if they failed on this they are putting too many eggs in one basket" / "I think this was more the supply chain that were not interested rather than the ODA selecting one person"
BOVIS LEND LEASE – RESPONDENT 1	"…you got to think about it in that it's in the shadow of Wembley, there were so many issues with Wembley with managing the project, a lot of contractors said why on earth would we divert time and resources and risk into a project like that when there is much other valuable work going on out there"
BOVIS LEND LEASE – RESPONDENT 1	"I think if I was a major contractor I would say, where is the risk? Where am I best placed to make money in the construction industry at the moment? A big risky job like the stadia for the Olympics or a nice big prestigious office block in the middle of town for somebody like Land Security, where I know I can put a premium on and its fairly safe money, its fairly robust design, that's lower risk than something like the Stadium, and I think that's what a lot of companies did" / "…why jeopardise and ongoing client like Land Securities or British Land by putting all of your resources into this job (Olympics) and shutting out a client that has got something like £50 billion worth of invest of the next 10 years, why would I upset a Land Security to do one job and maybe it goes all sour and pear-shaped, and my reputation goes down the pan because of one job and yet I have managed to maintain a client for the last 10 years and I want to keep with that client for the next 10 years."

BOVIS LEND LEASE – RESPONDENT 2	"limited depth of supply chain in the type of contractor that could deliver it, they are not many people that can do big stadia in this country"
BOVIS LEND LEASE – RESPONDENT 2	"I think most of the contractors turned around and they could see what the design was and they knew what the budget was and they said, can't do it"
BOVIS LEND LEASE – RESPONDENT 3	"I don't actually think they had a choice given the space of the market over the last 2 years, the market has been very over heated, there has been huge amount of construction work going on, particularly in the south-east, particularly in the city, and even going through a process of negotiations the ODA had difficulties finding people wanting to bit on certain buildings"
ODA	"I mean the stadium is a bit of a hybrid because there was only one player so you were in a position…where you only had one player and the option was Hobson if you're not careful but actually the player we had was the right player for the job." / "Team McAlpine are probably the leading players in the marketplace for building stadiums in terms of design…who have a history of building these things successfully currently within the market place. I mean they successfully delivered the Emirates Stadium. So the idea is to put a team together of people who have done this before and understood it and actually re-engineer it…"
ODA	"We couldn't get enough appetite in the market to bring anyone else to bear. I mean the Multiplex issue, the issue that was in the press at the time. It was a huge factor on this and basically we did have more than one player at PQQ stage, it only ever got to PQQ, there never was a tender document. It went to PQQ and basically out of PQQ no one qualified apart from team McAlpine…"

BAA	"It is likely to be a strategic decision, is it going to be more hassle than it's worth or are we going to get some sort of recognition at the end of the day for having worked on a successful project? I think one of the things that we sold to our supply chain on Terminal 5 was around making Terminal 5 not just a success for BAA but also if they were successful on T5, so we would be successful, they would be successful so therefore within the industry we would all look as though we had all done a great job so therefore it would be a good marketing opportunity for them as a business as well as us because they are the ones going for the continued building construction projects, and the supplier I was talking to around the Olympics was actually saying well I have got to weigh up at the moment whether I think I am going to get any value out of this project."
BAA	"…the other thing that is slightly going against the ODA is it's a one off thing. Where as you see for example BAA over the next 5 years we are spending £6 billion. So basically the supply chain is going to say to us you have got more work so we are probably going to perform a little bit better and make sure I do perform because I know there is continual work flow whereas on the Olympics they still want to be seen as delivering it well because obviously that goes well in their portfolio but you are not getting the actually repetition business and that sometimes can ago against you as a one off client in a way about driving performance and driving best price."
BAA	"…yes they are looking at Wembley and oh that didn't go too well and do we really fancy getting involved because it's going to be high profile."
Competitive Dialogue on Aquatic Centre	
ODA	"Yes, we did a big piece of market warming and we would have, yes, definitely liked more players to come and step forward. There were big issues to get round really. Again it was a venue and Multiplex was still ringing in the ears of people.

	Multiplex have stopped working in this country as a contractor on the back of Wembley, and the whole debacle with them, you know the FA and others made the market very nervous about sports related projects. So we had a hell of a lot of work."
BOVIS LEND LEASE – RESPONDENT 1	*"That sort of dialogue drives out whether you really have got a true partnering culture on board, a lot of main contractors and a lot of clients talk the talk, but very rarely put that into practice, they all revert to form after a while, when the going gets tough they all revert to lowest cost, quickest programme, whereas within partnering, you do true partnering and work closely with the supply chain, you need to get the cost right but you can get all sorts of benefit from that process"*
BOVIS LEND LEASE – RESPONDENT 1	*"I think that does come out of that dialogue process because it assumes that you always get the price and the programme right, put that to one side, we need to talk about the culture of the people in your business, are they going to work collaboratively with us to deliver the best possible project"*
BOVIS LEND LEASE – RESPONDENT 1	*"…at that scale, we are talking about a major project, we are not talking about some extension to somebody's house, at that scale of project I think that level of interview probably works better."*
Ferrovial	*"Did they do the right thing, I think so, yes because again, these people in the ODA, they are professionals, they know what the construction process is so if you just did a little study and said "okay, if we were to consult with everybody that we normally consult with and at the end of that consultation we were still going to come up with 3 people, then why not just do the straight negotiation with those 3 people, providing you can demonstrate that we are going to short circuit the process because of these reasons."*

Ferrovial	*"It is very important that you understand the people on the other side of the fence, how do they work, what is the market, how do they operate, what motivates them, you know that kind of thing. So, I think given the time, the pressures, the politics and all of that I think negotiation/competitive dialogue was the only option."*
BOVIS LEND LEASE – RESPONDENT 2	*"It is about people interacting with people rather than just doing a desktop study and say I pick that one (contractor) because his schedule of rates is lower, they actually sit across the table with the person that is going to deliver it and see how everybody interacts with everybody and ask questions such as how are you going to manage this and how are you going to manage that, what are the key areas for you and then its question and answer, but it is how they work together as team"*
BOVIS LEND LEASE – RESPONDENT 2	*"…is he a good person to do business with or is the kind of character that is going to keep everything to himself and not communicate, what is he like as a person."*
ODA	*"Competitive dialogue is very different, I mean, I ran the competitive dialogue for aquatics with the team, competitive dialogue, don't forget, takes you into a discussion of development of dialogue process before the ITT goes out. You can only negotiate if you've got an offer on the table. We never had an offer on the table compared to dialogue, you stop the dialogue and then you go out to tender." / "And once you go out to tender it's restricted so you can't be flexible about what you are offering. So in other words, you get your contractor, we got Balfour Beatty in the room and over a period of three months we developed our thinking with them independently, …, and then when we decided we had had enough and we had got enough to actually put three independent bespoke ITTs out, we stopped the dialogue and went into what is known as a restricted process."*

Form of contract: Target cost contract

	Why Target cost
BOVIS LEND LEASE – RESPONDENT 1	"…you got to remember all of that was happening in the shadow of Wembley, spiralling cost in Wembley, difficulties with the supply chain, it was overrunning, so the supply chain perceived this to be yet another major project and put the premium on because of that." / "cover the risk of the unknown and the difficulties associated with it and the profile of it…contractors looking at this job would perceive it to be similar as Wembley and want to price the risk accordingly and also put a premium on because it is an opportunity." / "I would perceive that this is exactly what happened & premium, premium, premium was added and the client decided in the end that they take the risk."
BOVIS LEND LEASE – RESPONDENT 2	"…they probably had all sorts of stuff buried in there from a technology point of view, that also pushes the price up, its the ODA, it's the Olympic Games, it's the public purse, they cough up the money because they have to have the games, they not going to not do it, so its another reason for putting another mark up on."
BOVIS LEND LEASE – RESPONDENT 2	"I think, because the design and build cost were coming in so high because McAlpine put so much contingency in there because, hey, I can see the pretty picture from the outside and that looks like it is going to be a lot of money and because it is very iconic, it is a one off, we have not done this design before, even though we done the Emirate Stadium, lets put a big contingency in there…"
BOVIS LEND LEASE – RESPONDENT 3	"with very large infrastructure projects, whereby, essentially it is never been done before, so to try to get a lump sum price from a contractor when the design does not exits or it exist to such an extent that it is meaningless, either means, that the contractor is going to charge an enormous amount of premium for taking the risk or he is going to leave you with the risk

	and is not going to take it at all, and just price what is on drawing, which is next to nothing, you then end up in a situation where you constantly battling, whereas, you can go with target cost and you got to have a set of rules, you stand half a chance to get to a sensible answer and not getting ripped off."
BOVIS LEND LEASE – RESPONDENT 1	"…if you want an early start but you have not got a clearly defined design you can at least engage with the trade contractor, it ring fences profit and overheads and that sort of stuff and then move onto the next stage to firm up on the cost, so it allows you to make a start to move forward without having to finalise the design, that is the advantage of a target cost contract, and then you are trying to overcome the variables between companies and establish that, all being equal, they will work with you to develop the target cost and the design." / "…you are using the experience of the trade contractor to validate the price."
ODA	"He has got to demonstrate that he can spend all the money you've got for him to spend. With a fixed price lump sum contract you've got a risk adverse contractor, in a single negotiation with no commercial leverage, he'll name a number, if he only spends fifty percent of that number that's marvellous. With target cost contracts you only pay him what he is due and you will then get a pain/gain share model. So therefore you mitigate the risk of paying him all that risk premium."
Ferrovial	"…so, if I am a contractor building this I know what the target is and if I beat that target for example, if I come underneath that target there might be an incentive for me to get more money, so what I am going to do is make sure I lay the bricks once, or I only do something once, I do not duplicate, I get everything right first time. So, therefore, if I do that and it actually motivates me if I put my contractor hat on to deliver this facility in the most efficient manner because I want that money by the way, because I'm here for profit so because I think like that, I'm going to do that, now if you don't convert it over to the clients side the client is happy because he knows that if he has chosen the best constructor then he is not taking

	any of the risk anyway because the contractor wants the money. If he breaches the target there is no money for the contractor, additional income."
ODA	"Arguably, if you had two identical projects, on an identical way and identical project managers, would you have any difference between a target cost outcome and a fixed price outcome. I put it to you, maybe not. But all you do is you get there without this sort of adversarial relationship that the more traditional fixed price job encourages. That's kind of the argument for target costs."
ODA	"The advantages are quite significant when you come to looking at change. What you've got is, you've got a contractor who is incentivised to actually work with you, to continue working and actually delivering what your requirements are in real time, rather than not do work because he hasn't had it approved under variation instruction."
ODA	"On the basis that fixing a fee and then reducing the actual costs to a minimum, so in other words it makes it more attractive to the supply chain in terms of, he is making more profit on lower turnover"
BOVIS LEND LEASE – RESPONDENT 3	"D&B and Target cost is an unusual choice, the NEC is perceived as a fairer contract to both the contractor and the client and is less client orientated, most clients in that situation would go and get a bespoke contract that puts more risk on the contractor."
ODA	"...because every compensation event, unlike the JCTs the ICEs and the like, every compensation event has consideration for time, whether it needs to adjust the programme or not, so in other words NEC for a fundamental change that might be prescribed by the client, the change must have time and cost considered within it. And what you will find with Government

	projects where they have got into trouble is there has been no transparency in why these changes have been agreed to. NEC has got kind of what we call embedded corporate governance. It's compensation event, early warning structure whereby risks are identified and mitigated, costs are mitigated and then compensation events to adjust the target, to adjust the accepted programme to allow for modifications to acquire this to go forward. So in essence we start with a contract that's actually very good at mitigating risks if used appropriately. But of course the risk is that we don't use it appropriately."
BOVIS LEND LEASE – RESPONDENT 2	"Yes, Target cost contracts can be made successful."
	Target Cost – pitfalls
ODA	"There are always issues about resolving the target and what the target is. Keeping the target current and making sure that you don't lose sight of the target and overpay the contractor towards the end of the process. Because you can easily have the situation where you have paid him too much where you re-adjust the target with the last few compensation events before the contract's resolved, you end up finding you've paid him too much already and he owes you some money."
BOVIS LEND LEASE – RESPONDENT 3	"…you can make it work if all checks and balances are in place and you got to have faith in the people that you are contracting with and they understand what you are saying and they are seeing this as a huge opportunity to rip the client off and charge you a fortune, so, it is very contractually commercially, but it can work, but it is not easy and you have got to have the right people in place to do it"
BAA	"I mean again target costs are fine provided you administer the contract robustly and that's where you know post-contract management of it. If you are not managing change and valuation of compensation events etc that is where it can run away from you."

BOVIS LEND LEASE – RESPONDENT 3	"…you can get into the sort of arguments, what is included in the cost and what is not included, what is a compensation event, who delayed the design, you as a contractor put the design forward, the client might look at it for too long, and think, it is not what I think it is, who pays for that, you have to make sure that the drawings comply with the brief, in essence, you can push all those issues on to the contractor, but you as client have to make sure that you have somebody on your side to sign all those things off and actually understand what you (as client) are trying to achieve."

Overcome the pitfalls

BOVIS LEND LEASE – RESPONDENT 1	"Using CLM to manage a target cost option does not cost the ODA any more money, the people are there"
BOVIS LEND LEASE – RESPONDENT 3	"…clarity in as much as the clarity of the brief, clarity in terms of the commercial arrangements, in terms of what is allowed and what is disallowed and you can go through each one of the categories and it needs to be absolutely clear, it also needs to be fair, so that both the contractor and the ODA stand a chance of actually getting to the end in a sensible working relationship, in a partnering kind of way, lots of people talk about partnering together with target cost and only really works when the deal in itself works and the contractor feels that they are making money and they probably making more money then they expected, which is always good, and the client feels that the building is being produced on time and to the quality that was required, and when you get to that equilibrium it works really well and everybody will sing the praises of this contract." / "At T5 this type of contract worked really well because the contractor made money and the client got a building when he expected it, to a budget that he also expected"
ODA	"…you need to be pro-active as a project manager in terms of costs. And also you need a tool to manage your contracts. In other words, target costs NEC contracts don't get managed by paper, anything over about fifteen million you are going to struggle." "you need a tool like BIW or a number of other proprietary NEC contract administration tools out there. You

	need to have one because you switch on your PC in the morning, there's the project manager for saving to see how many outstanding compensation events you've got, how many quotations you've got due, how many early warnings are in, how you're going to mitigate them, what the current target price is, the lot." [And do you have that system?] "We do have that, yes. We have to have it."
BAA	"...it is also being very rigid about the target cost, you as a client, saying I can only afford that. In other words somebody coming out and saying I can only like we said we can only afford £4.3 billion."